# Statistical Methods for Pharmaceutical Research Planning

# STATISTICS: Textbooks and Monographs

*A SERIES EDITED BY*

D. B. OWEN, Coordinating Editor

*Department of Statistics*
*Southern Methodist University*
*Dallas, Texas*

## R. G. CORNELL, Associate Editor, Biostatistics

*Department of Biostatistics*
*University of Michigan*
*Ann Arbor, Michigan*

## OTHER VOLUMES IN PREPARATION

# Statistical Methods for Pharmaceutical Research Planning

## STEN W. BERGMAN

Mortgage Finance Department
Salomon Brothers, Inc.
New York, New York

## JOHN C. GITTINS

Keble College
Oxford University
Oxford, England

CRC Press
Taylor & Francis Group
Boca Raton London New York

CRC Press is an imprint of the
Taylor & Francis Group, an **informa** business

First published 1985 by Marcel Dekker Inc.

Published 2019 by CRC Press
Taylor & Francis Group
6000 Broken Sound Parkway NW, Suite 300
Boca Raton, FL 33487-2742

© 1985 by Taylor & Francis Group, LLC
CRC Press is an imprint of Taylor & Francis Group, an Informa business

First issued in paperback 2019

No claim to original U.S. Government works

ISBN-13: 978-0-367-45166-0 (pbk)
ISBN-13: 978-0-8247-7146-1 (hbk)

Visit the Taylor & Francis Web site at
http://www.taylorandfrancis.com

and the CRC Press Web site at
http://www.crcpress.com

Library of Congress Cataloging-in-Publication Data

Bergman, Sten W.
    Statistical methods for pharmaceutical research planning.

        (Statistics, textbooks and monographs ; v. 67)
        Includes bibliographies and index.
        1. Pharmacy—Research—Statistical methods.
I. Gittins, John C., [date]. II. Title.
III. Series.
RS122.B45  1985        615'.1'072        85-16097
ISBN 0-8247-7146-X

# PREFACE

Each year an oak tree produces hundreds of acorns. Only a few of these germinate and take root, and only once in a hundred years or more does one of them survive to become a mature tree. Pharmaceutical and agro-chemical research follows a similar pattern. A laboratory may synthesize hundreds of compounds per year. These are then subjected to relatively inexpensive screening tests. The most promising compounds are then tested on animals, or in field trials, and occasionally one does well enough to be developed and marketed as a commercial product.

Statistical methods are relevant throughout this process, but in this book our chief concern is with the early speculative phase of research, characterized by large numbers of test compounds. The emphasis is also on methods which impinge more or less directly on the decisions that are made during the course of this kind of research, rather than just on those that are purely informative. Four decision-making areas are considered: the selection of compounds for testing; the design of a screen; the sequencing and allocation of resources between a set of related screens; and the allocation of effort between different projects.

The four chapters of the book deal with these areas in turn. The relevant literature is reviewed, and the principal methods are described in some detail. To these we have added some of our own, including some not previously published. Chapters 1, 2, and 4 all begin with quite long introductions. They summarize and evaluate the entire contents of these chapters, and are less technical than remaining sections of the chapters; hopefully, the busy manager will find that they tell him what he needs to know. Each of these chapters is followed by an extensive list of references, many of which are not specifically mentioned in the text. The references in Chapter 2 give a reasonably comprehensive coverage of the literature, and do not fall far short of that for Chapter 4. The potential reference list for Chapter 1 is enormous, and the list given reflects the emphases of the chapter itself, with only a representative selection in some areas. The book as a whole is aimed at all scientists engaged in the speculative end of new-product chemical research, particularly those with a responsibility for any kind of planning, together with the statisticians and operational researchers who advise them.

The final postscript considers the adequacy of existing statistical procedures in each of the four areas and suggests some extensions and improvements. The temptation for us as academics is to put forward objections to the methods that have been proposed so far, and to invite our industrial friends to supply us with funds so that we can develop better ones. However, we should like to emphasize that, while there is always room for improvement, there are procedures already available for three of the four areas, the exception being the organization of sequences of screens. The material on offer seems to justify greater use of statistically based methods than has so far occurred. For most problems it should be possible to find something in the published literature which serves as a starting point for working out a useful procedure.

We should like to thank those colleagues in five different industrial laboratories, four of them British, who discussed current practice with Sten Bergman when he visited them during the preparation of this material. Most of all we should like to thank Arend Heyting, of Duphar B. V., Weesp, Holland, whose firm sponsored the entire exercise. We should also like to thank Val Willoughby and Gay Breuler for typing, respectively, chapter texts and the reference lists that accompany them. Finally, we express our thanks to our wives, Carmen Mansourian and Elizabeth Gittins, for their continued encouragement as well as valuable assistance during the production phase of this book. In gratitude, we dedicate this work to them.

Feedback would be very welcome. Inquiries about software for the DAI and RESPRO procedures (see Sections 1.4.5 and 4.9) should be sent to John C. Gittins (Mathematical Institute, 24-29 St. Giles, Oxford OX1 3LB, England), while further information about the fixed-point procedures of Section 2.6s and 3.2 should be obtained from Sten W. Bergman (Mortgage Finance, Salomon Brothers, Inc., One New York Plaza, New York, N.Y. 10004).

*Sten W. Bergman*
*John C. Gittins*

# CONTENTS

# Statistical Methods for Pharmaceutical Research Planning

# 1

# QUANTITATIVE STRUCTURE-ACTIVITY RELATIONSHIPS (QSAR)

## 1.1. INTRODUCTION

The notion that the biological activity of a compound is a function of its chemical structure was initially proposed by Crum-Brown and Fraser (1869). Implicit in this notion is the possibility of scientifically designing compounds for specific diseases. Real progress toward discovering these structure-activity relationships (SAR) between the physicochemical properties of compounds and the observed biological response in an organism did not, however, occur until the early 1960s, when Hansch and coworkers (see, e.g., Hansch, 1971) observed a definite correlation between certain physicochemical characteristics of a set of congeners* and their biological activity. This work marked the beginning of a large and ever-growing literature on quantitative structure-activity relationships (QSAR). Presently, there are many hundreds of papers on this subject, referencing scores of techniques.

In the pages that follow this introduction, a detailed presentation is given of some selected QSAR methods. We have chosen to review two of the most important areas in Sections 1.2 and 1.3, where we discuss regression-related methods and classification. They define two areas for which the problem is well formulated and there are available several interesting techniques. Some of the individual methods discussed in detail in these sections are

---

*That is, structurally similar chemical compounds.

well known within the pharmaceutical literature. Others are not, and among these are some that warrant more attention in view of their potential usefulness.

In addition, five other techniques are presented in Section 1.4. Two are based on cluster analysis, two on the identification of key variables influencing activity, and one on the assignment of testing resources among competing groups of compounds. Four of these are either unknown or not well known within the QSAR literature. We have chosen to draw attention to these five because of either their potential usefulness or their inherent methodological interest.

In the remainder of this introduction, a nontechnical overview is given of these selected methods. Simultaneously, a guide is provided to some other aspects of the QSAR literature.

### 1.1.1. Regression and Related Methods

One of the most versatile techniques in statistics, with wide use already in QSAR studies, is multiple linear regression. Hansch and coworkers (see, e.g., Hansch, 1971) have used regression techniques to predict the activity of a compound belonging to a set of congeners from its electronic, steric, and hydrophobic characteristics. Free and Wilson (1964) also used a type of regression equation to determine the relative contribution that the presence of a specific substituent in a given site has toward overall activity for a compound belonging to a set of congeners.

The two models noted above are examples of the general linear model discussed in Section 1.2, which encompasses regression, analysis of variance, and analysis of covariance. The basic assumption is that except possibly for random variation (conventionally, though not always accurately, referred to as error), the activity of a compound is linearly related to a set of physicochemical features possessed by the compound. This linear relationship is assumed to hold for all compounds under consideration, and consequently for each compound i one has

$$Y_i = \sum_j \beta_j x_{ij} + \varepsilon_i$$

where $Y_i$ is the observed activity of the ith compound, $x_{ij}$ the observed value of the jth feature, $\varepsilon_i$ an error term, and $\beta_1, \ldots, \beta_m$ a collection of coefficients that do not depend on i.

Typically, the coefficients $\beta_1, \ldots, \beta_m$ are unknown, and an initial objective is to estimate their values and use the resulting equation to predict the activity of compounds not yet tested. To obtain unbiased estimators for them with desirable statistical properties, it is sufficient that the independent variables $x_{ij}$ be measured without error, and the the error terms $\varepsilon_i$ be independent and identically distributed. It is not required that the common distribution be the normal distribution.

To go further and test with exact significance levels various hypotheses regarding the coefficients, stronger assumptions are needed. It is in this context that the assumption of normality is convenient. Some aspects of hypothesis testing are reviewed in Section 1.2.1.4, and additional information may be obtained from several standard textbooks (e.g., Draper and Smith, 1966 and Seber, 1977). To test the adequacy of the linear model itself one needs an estimate of the error terms which is independent of the model. This can be achieved by repeated observations on the activity of the compounds being tested.

The power of regression techniques arises from their ability adequately to model complex relationships between the activity of a compound and its physicochemical attributes. In particular, the final set of independent variables $x_{i1}, \ldots, x_{in}$ may be derived from the original features by taking powers and other transforms of their values, and hence the linear relationship $Y_i = \Sigma_j \beta_j x_{ij}$ may be made to represent to an adequate degree virtually any functional relationship that exists between activity and these features. Particularly useful in this respect is the use of indicator variables to represent the presence or absence of certain features, either singly or jointly, which appear to have an impact upon activity.

In the QSAR literature, linear regression has been restricted primarily to analyzing a set of congeners. However, from a purely statistical point of view there is no necessity for such a restriction, and linear regression may be used to model the structure of large data sets consisting of unrelated compounds. Later in this chapter some examples are given of how this may be done using indicator variables. Related techniques may be found in Draper and Smith (1966) or Seber (1977). The success of any such attempt, of course, ultimately depends on how useful the resulting equations are for identifying potentially active compounds.

Another advantage of the regression methodology is the ubiquitous availability of computer software. Virtually every statistical package contains an easy to use regression subroutine.

An increasingly common practice is to use regression techniques to discover significant predicting variables. For example, given a set of dependent variables $Y_i$ (i = 1,...,n) and a large set of independent variables $x_{ij}$ (i = 1,...,n; j = 1,...,m) the objective is to discover a parsimonious subset of the independent variables which will adequately predict the $Y_i$ values. It is well known (e.g., Kendall, 1975) that a simple test of the significance of each variable in the final regression equation will not in general yield a suitable subset.

In Section 1.2.2 three techniques for finding such a subset are reviewed. The simplest of these is the complete subset approach, in which a regression equation is fitted to every subset of the independent variables. The software package BMD contains a subroutine that performs this task. However, with more than 11 or 12 independent variables, the number of regressions becomes intolerably large. A widely used alternative is stepwise regression. Here, variables are entered or removed from the regression equation in a stepwise manner until a final subset is obtained. In the Efroymson approach (1960) this final subset is the first subset S such that in the regression equation based on S, all the variables are statistically significant, while in the regression equations based on S and any additional variable, this additional variable is statistically insignificant.

As has been pointed out by Topliss and Edwards (1979), one of the dangers of the stepwise approach is the heightened chance of obtaining a spuriously good fit. Using simulation studies, they provide rules that will permit the user to guard against a high multiple correlation coefficient occurring by chance only. The most useful of these rules are reproduced in Section 1.2.2.3. Another method to guard against spurious results is the use of the so-called jackknife, whereby the data are split into two parts, the first being used to find a good subset and the second to confirm the goodness of fit.

It has been suggested by some authors that the statistical technique known as principal components analysis may be used to identify a good subset. In an article by Martin and Paras (1979) this method is used on the explanatory variables alone to identify the essential dimensionality of the space defined by these variables. The principle here is that collinearities in the independent variables imply that the information contained in the total set can be recovered from a smaller subset. However, a danger in this approach is that the collinearities are never exact, and essential information concerning the dependent variable may be hidden in these near collinearities. Furthermore, the variables that are finally selected may be totally irrelevant for prediction purposes. These difficulties arise because the relationship between the independent variables and the dependent variables is not examined in this type of principal component analysis.

An alternative method has been suggested by Hawkins (1973) and it is reviewed in Section 1.2.2.4. Here a type of principal component analysis is performed on all the variables, including the dependent variable. The principle is now one of finding a subspace that contains regression equations all of which lie close to the best regression equation. Some of these equations may depend on only a few of the original independent variables, and thus identify a good subset. The technique is interactive, and it permits the user to find and select a subset not only on the basis of a good fit but also using other criteria, such as cost and ease of measurement. As such it appears to be a useful and promising technique. Unfortunately, however, the software is not yet readily available.

When the dependent variable is binary, simply indicating the presence or absence of a certain kind of activity, the results obtained by the standard regression model are often difficult to interpret. An alternative in such circumstances is the logistic regression model, in which it is assumed that activity occurs with unknown probability P. The objective is to estimate P under the assumption that the logarithm of the odds ratio is a linear function of the features, that is,

$$\log \frac{P}{1-P} = \sum_{i=1}^{m} \beta_i x_i$$

By estimating the unknown coefficients $\beta_1, \ldots, \beta_m$ by $\hat{\beta}_1, \ldots, \hat{\beta}_m$, one then obtains an estimate $\hat{P}$ for P, which is interpreted as the estimated probability that a compound with features $x_1, \ldots, x_m$ will be active.

This model has not yet been widely used in pharmaceutical research but promises to be useful in handling data characterized by a binary dependent variable. BMD includes a stepwise regression subroutine for determining the coefficients in the logistic model.

### 1.1.2. Classification

The second area discussed in detail later in this chapter is classification or, as it is sometimes alternatively called, discrimination. (In the QSAR literature, the term "pattern recognition" is also used for the subject matter at hand; however, this is a very broad term, encompassing many aspects other than classification.) The problem as defined here is one of using the information about the features of two correctly classified groups of compounds to classify a new compound on the basis of its features as belonging to one of these two groups. Typically, the two alternative classifications are active and inactive, and the principle of classification is that if a compound has features more characteristic of actives than of inactives, it will be classified as active. The techniques in use vary according to how this principle is made precise.

The simplest of the methods discussed is the k-nearest-neighbors method. A compound is classified as being active if a majority of its k nearest neighbors are active. Both empirical and theoretical evidence show that this simple rule gives good results. Its main disadvantage is that as the training set grows large so does the computational effort in determining which compounds are the k nearest neighbors.

Another nonparametric method is the linear learning machine. It is an algorithm for finding a linear function of the feature that takes larger values for compounds belonging to one of the training sets than for any of those in the other training set. This technique has received considerable attention in the QSAR literature (see, e.g., Stuper et al., 1979). However, its usefulness is in doubt. First, the linear function found is the first linear function encountered with separates the training sets. It is not the only such function, and may well not be the best one for predicting the activity of other compounds. Second, there is no a priori reason that the training sets should be linearly separable. Typically, they will not be. Third, if the training sets are not linearly separable, the standard algorithm never terminates. The last defect may be overcome by detecting cycling when it occurs, but then several passes through the data will be required. This may be expensive.

A different set of classification procedures is obtained through the decision-theoretic framework of Wald. It is assumed that there are costs associated with misclassifying compounds. The objective is one of finding a decision rule that minimizes the expected cost of misclassification. To get a handle on this problem, one may postulate the existence of two distributions for the features one is apt to observe for active and inactive compounds. One may then show that the optimal decision rule can be described in terms of these distributions. Consequently, it is sufficient to seek procedures that estimate them.

The first decision-theoretic procedure discussed in Section 1.3.4 is the Fisher linear discriminant function. It assumes that the two unknown distribution functions are multivariate normal with unknown means and an unknown but common covariance matrix. These are then estimated from the data by the maximum likelihood method.

These assumptions are somewhat restrictive, but the computation of the optimal decision rule is relatively easy with commercially available software (e.g., BMD; see Dixon, 1967). An immediate generalization may be obtained by permitting the covariances to be unequal.

Day and Kerridge (1967), starting from a further generalization of the multivariate normal model, have suggested the use of logistic distributions of the form

$$\frac{\exp(y^T C + d)}{1 + \exp(y^T C + d)}$$

where y is a feature vector, C an unknown column vector, and d an unknown constant. The problem is one of finding the maximum likelihood estimates of C and d for actives and inactives, respectively. This may be done by quadratic hill-climbing methods, but a complete program is not commercially available. Anderson (1972, 1974, 1975), in a series of papers, has further developed this model.

There are two advantages to the logistic model. First, the parametric assumptions, as argued by Day and Kerridge, are relatively weak. Second, if the training sets are linearly separable, then maximizing the likelihood functions will inevitably lead to a separating linear function. Thus the logistic model provides the benefits of the linear learning machine without some of the accompanying pitfalls. As yet there is no reference to the logistic model in the QSAR literature.

Even more promising are the kernel-based procedures of Habbema et al. (1974) and Aitchison and Aitken (1976). These, too, are unknown in the QSAR literature. While conceptually belonging to the decision-theoretic approach to classification, the assumptions are so weak as to make them virtually nonparametric, and quite similar in operation to the k-nearest-neighbors scheme. The idea is one of distributing a small amount of probability mass about each observed point, and then summing to register the amount of mass accumulated at each point in the feature space. This summing has the effect of creating hills and valleys describing the distribution functions for the features as observed for each training set. As is readily seen, the method is essentially nonparametric. Furthermore, it may be used with both continuous and binary features. In the example given by Aitchison and Aitken, the procedure was able to classify correctly all 41 unknowns on the basis of two training sets consisting of 40 actives and 37 inactives, respectively. (The actives in this context were patients with a specific disease, and the features were the presence or absence of individual symptoms in a set of possible symptoms.)

### 1.1.3. Cluster Analysis

Cluster analysis has also been used in QSAR studies. The two algorithms discussed in detail in Sections 1.4.1 and 1.4.2 are directed toward discovering unusual clusters of active compounds in a field of inactive compounds. The first is a well-known algorithm by Harrison (1968) which has been successfully implemented by Imperial Chemical Industries to discover new lead compounds. Based on the assumption that attributes are independently distributed among the compounds, a similarity measure is developed which has the property that compounds sharing k rare attributes are considered more similar than ones that share k common attributes. Unusual clusters of active compounds are then found by determining for all j the probability that of the j nearest neighbors to an active compound i of them are active.

The assumption that the attributes are independently distributed among the compounds is strong and will, in practice, not be satisfied. However, the derived similarity measure may still be useful. Further work on the Harrison model has been published by White and Lewinson (1977).

A clustering algorithm for the same problem that Harrison considers has recently been developed by one of the present authors (Bergman, 1982). It assumes that there exists a similarity measure between compounds. This measure may be the one derived by Harrison, or it may be any other similarity measure thought to be appropriate. The graph in which every pair of compounds is joined by a series of links, and these links are chosen so that their total "length" in terms of the similarity measure is minimized, is then computed, and clusters are formed by removing all those links connecting adjacent active and inactive compounds. All large active clusters are then examined for a common underlying explaining cause. The advantages of this procedure are that it is nonparametric; it is computationally effi-

cient, being able to handle large data sets; and a significance
test may be derived for testing the hypothesis that actives and inactives do not have different distributions of features.

The more traditional clustering algorithms, in which the objective is simply to find clusters of compounds in the feature
space without regard to their activity, have also been found useful in pharmaceutical research. Their principal application is one
of reducing the number of compounds to be tested, thus permitting the chemist to select representative compounds from clusters
of similar compounds (see, e.g., Martin and Paras, 1979). These
techniques are not reviewed here, but some of the principal
methods are k-means, single linkage, nonlinear mapping, and
Andrew's method. The later two are graphical. Further information on these and many other clustering algorithms may be found
in the textbooks by Everitt (1974) and Hartigan (1975).

### 1.1.4. Key Variables

Another area of the QSAR literature is represented by techniques
which seek to discover the key variables that influence activity.
Two methods with applications to binary feature data are reviewed
in detail in Sections 1.4.3 and 1.4.4.

The first of these, CHAID (chi-squared AID) by Kass (1979),
is a derivative of Morgan and Sonquist's automatic interaction detection (AID) algorithm. The idea here is to seek that independent variable which is best able to predict the binary dependent
variable. If its prediction ability is significant, the data are sub-
divided into two subsets according to the two values of this independent variable. Thereafter the process is repeated on each of
these two subsets separately, generating successive generations
of subsets, until finally none of the remaining variables for partitioning a subset is able to significantly predict the dependent
variable. CHAID appears to be a useful tool for identifying the
features that influence activity. It has not yet appeared in the
QSAR literature. Unfortunately, no commercial software is available, but it does not seem too difficult to program.

An interesting but complex key model has been proposed by
Brown (1970) for determining the best model for an observed response. It is assumed that molecules in a series of congeners
share a set of major sites, and that the molecules differ by the
substituents occurring at these sites. The task is to determine
which sites and which substituents in these sites are instrumental
in producing the observed degree of activity (i.e., active or inactive), given that the recorded response does not always reflect

the true activity of the molecule. Each such explaining combination of sites and substituents is called a model. For the observed set of results the method calculates the probabilities that these were caused by each of the possible models, thereby providing a basis for choosing between the models. The models with the highest posterior probability are then the most likely models for explaining the observed response.

### 1.1.5. Allocation of Research Effort

A vital aspect of successful pharmaceutical research is the appropriate allocation of research effort. This topic may include such diverse aspects as project selection, screen design, and lead optimization. Chapter 4 discusses the first of these in detail, while Chapters 2 and 3 explore elements of the second. However, for the biochemist the immediate question is often simply which compound to test next. Some QSAR methods attempt to give a more-or-less direct answer.

The key model by Brown mentioned in Section 1.4 provides some useful information. In Section 1.4.6 a procedure proposed by Gittins and Jones (1974a) is described. This suggests that dynamic allocation indices (DAIs) be used to help decide from which group of compounds the next observation should be taken. The DAI for a group of compounds depends on the test results for the compounds from the group that have already been screened. On the assumption of a homogeneous distribution of activity within each group of compounds, the DAIs are calculated in such a way that a policy which minimizes the expected cost of finding a compound with a specified level of activity is one which always selects a compound for testing from the group whose DAI value is currently largest.

Except when compounds are randomly selected, the homogeneity assumption is at best only approximately satisfied. However, with care, useful results may still be obtained when it does not hold. Most obviously, the various groups of compounds might correspond to different leads. The procedure has also been used when each group corresponded to a particular site at which the lead compound, which was the same for every group, had been modified. Although so far used only on an experimental basis, the indications are that the DAI procedure is capable of achieving substantial reductions in the number of compounds that must be tested in the search for interesting levels of activity.

A quite different approach is to attempt to find the maximum of the response surface that defines activity. This response surface may be regarded as a function $f(x_1, \ldots, x_m)$ of the physico-chemical features $x_1, \ldots, x_m$ one measures for each compound, and whenever one measures the activity of a compound with a given set of features $(x_1, \ldots, x_m)$ one observes $f(x_1, \ldots, x_m) + \varepsilon$, where $\varepsilon$ is an error term due to biological variation. The objective in this setting is to determine those values $(x_1^*, \ldots, x_m^*)$ that maximize $f(x_1, \ldots, x_m)$.

Two different methodologies exist in the literature for finding these maximizing values. First, one may determine the response surface's general configuration by evaluating f over a grid of different values. Thereafter a quadratic function is fitted over the relevant data points, which is then solved analytically for the maximum. The parabolic Hansch model has been used in this manner, and a further discussion of it may be found in Section 1.2.2.

Second, one may proceed to find the maximum of f without determining the general form of the response surface. The Darvas simplex method (1974) illustrates this approach. It is essentially a hill-climbing algorithm that moves in the direction of steepest ascent. A related technique is stochastic approximation, utilizing versions of the Robbins-Monro algorithm (1951) for finding a maximum under uncertainty.

An advantage of the hill-climbing algorithms is that they find a maximum with relatively few observations; the disadvantage is that this maximum may only be a local, rather than a global maximum. This danger may be reduced by restarting the algorithm with different initial compounds.

A question relevant to the foregoing topic is whether it is better to take repeated observations to reduce error or to observe a new point. By simulation Spendly et al. (1962) have shown that it is generally more efficient to take an observation on a new point rather than a repeat on an old point.

## 1.2. REGRESSION AND RELATED METHODS

One of the most versatile techniques in statistics, with already wide use in QSAR studies, is multiple linear regression. In this section we discuss some aspects of regression.

First, the general linear model is introduced, and it is shown that both the Hansch and Free-Wilson models are examples of this model. Some suggestions are then made as to how the general linear model may be usefully extended to other settings. Thereafter the assumptions underlying estimation and hypothesis testing are clarified, and a test is given for determining the adequacy of the proposed model.

Second, the question of variable selection is partly reviewed. This includes subset selection and stepwise regression. A paper by Topliss and Edwards (1979) casts some light on the true significance of the multiple correlation coefficient in a stepwise regression context. Finally, an interesting technique by Hawkins (1973) for finding a collection of good subsets is described.

The last portion of this section discusses logistic regression and its usefulness for handling a binary dependent variable used to describe a compound as being either active or inactive.

### 1.2.1. General Linear Model

#### 1.2.1.1. The Model

In the general linear model it is assumed that there are n independent random variables $Y_1, \ldots, Y_n$, each of which is linearly related to a corresponding set of observations. In particular, if $x_{i1}, \ldots, x_{im}$ are the observations associated with the random variable $Y_i$, it is assumed that there exist coefficients $\beta_1, \ldots, \beta_m$, not depending upon i, such that

$$Y_i = \beta_1 x_{i1} + \beta_2 x_{i2} + \cdots + \beta_m x_{im} + \varepsilon_i$$

where $\varepsilon_i$ is an error term. The exact assumptions that we will wish to make regarding the error terms $\varepsilon_i$, $i = 1, \ldots, n$, will depend on what we are trying to do. Initially, at least, we shall not need to assume more than that the $\varepsilon_i$'s are independent random variables with mean 0 and unknown but common variance $\sigma^2$. This is sufficient for obtaining least-squares estimators for the unknown coefficients $\beta_1, \ldots, \beta_m$ with certain desirable statistical properties. However, when we wish to obtain exact distributions for testing the validity of certain hypotheses regarding our estimators, we shall need stronger assumptions, such as that the error terms have a normal distribution.

The general linear model may be illustrated with two well-known models found in the pharmaceutical literature. First, Hansch (see, e.g., Hansch, 1971) has postulated that biological activity in a set of congeners is linearly related to a set of physical parameters. In particular, the parabolic Hansch equation is

$$-\log c_i = a + b\pi_i - c\pi_i^2 + d\sigma_i \quad (i = 1, \ldots, n)$$

where $c_i$ is the concentration required of compound i in the series to produce a standard biological response; $\pi_i$ the pi substituent constant, measuring the ith compound's hydrophobic properties; and $\sigma_i$ the sigma substituent constant, measuring its electronic properties. Here a, b, c, and d are a set of unknown parameters to be estimated for the series.

The Hansch equation is easily written in term of the general linear model by setting $Y_i = \log(1/c_i)$, $x_{i1} = 1$, $x_{i2} = \pi_i$, $x_{i3} = \pi_i^2$, $x_{i4} = \sigma_i$, $\beta_1 = a$, $\beta_2 = b$, $\beta_3 = c$, and $\beta_4 = d$. From this example we note that the variables $x_{ij}$ may be continuous, discrete, or a mixture of both. Note also that the assumption of linearity refers to the parameters $\beta_1, \ldots, \beta_m$, not to the variables $x_1, \ldots, x_m$. In particular, we have here $Y_i = \log(1/c_i)$, $x_{i2} = \pi_i$, and $x_{i3} = \pi_i^2$.

Our second example is the Free-Wilson model (1964) for a set of congeners. It is assumed that every time a particular substituent group appears in the same place in the common base molecule, it will play a constant additive role toward determining the overall biological activity. For example, suppose that there are two substitution sites A and B in the base molecule, where A may be one of three different substituents ($a_1$, $a_2$, or $a_3$) and B may be one of two ($b_1$ or $b_2$). Then if biological activity is measured by $\log(1/c)$ we have the model

$$-\log c_i = \mu + \alpha_i + \beta_i + \varepsilon_i \quad (i = 1, \ldots, n)$$

where $\mu$ is a constant and $\alpha_i$ and $\beta_i$ are the contributions of the ith compound's A substituent and B substituent, respectively.

Hence the appropriately formulated linear model is the following:

$$Y_i = \gamma_1 x_{i1} + \gamma_2 x_{i2} + \gamma_3 x_{i3} + \gamma_4 x_{i4} + \gamma_5 x_{i5} + \gamma_6 x_{i6} + \varepsilon_i \quad (1)$$

where

$$x_{i1} = 1$$

$$x_{i2} = \begin{cases} 1 & \text{if } A_i = a_1 \\ 0 & \text{otherwise} \end{cases}$$

$$x_{i3} = \begin{cases} 1 & \text{if } A_i = a_2 \\ 0 & \text{otherwise} \end{cases}$$

$$x_{i4} = \begin{cases} 1 & \text{if } A_i = a_3 \\ 0 & \text{otherwise} \end{cases}$$

$$x_{i5} = \begin{cases} 1 & \text{if } B_i = b_1 \\ 0 & \text{otherwise} \end{cases}$$

$$x_{i6} = \begin{cases} 1 & \text{if } B_i = b_2 \\ 0 & \text{otherwise} \end{cases}$$

Consequently, $\gamma_2$, for example, represents the contribution of the presence of $a_1$ in site A to overall activity. We may also note that in this model the variables are either constant or binary.

More complex situations may also be modeled within this framework. For example, suppose that in the Free-Wilson setting it is believed that a specific combination of substituents at the A and B sites influence activity in a nonadditive fashion. Then an interaction term representing the presence or absence of this combination could be added. For example, if there were thought to be some special joint effect when $A = a_1$ and $B = b_1$, the model could be changed to reflect this by writing

$$Y_i = \gamma_1 x_{i1} + \gamma_2 x_{i2} + \gamma_3 x_{i3} + \gamma_4 x_{i4} + \gamma_5 x_{i5} + \gamma_6 x_{i6} + \gamma_7 x_{i7} + \varepsilon_i$$

where $x_{i7}$ is a new term defined by

$$x_{i7} = \begin{cases} 1 & \text{if } A = a_1 \text{ and } B = b_1 \\ 0 & \text{otherwise} \end{cases}$$

As a second illustration, suppose that in the Free-Wilson setting there were three quite different base molecules, each with sites A and B and each with an accompanying series of congeners. Typically, these series would be analyzed separately. However, if it were thought that the relative contributions of the substituents in these sites were the same once the general level of activity associated with the base molecule was accounted for, these three series could be analyzed simultaneously by introducing a variable indicating the series type. Thus the one-series model displayed in (1) is extended to become

$$Y_i = \gamma_1 x_{i1} + \gamma_2 x_{i2} + \gamma_3 x_{i3} + \gamma_4 x_{i4} + \gamma_5 x_{i5}$$
$$+ \gamma_6 x_{i6} + \gamma_7 x_{i7} + \gamma_8 x_{i8} + \varepsilon_i$$

where $x_{i2}$, $x_{i3}$, $x_{i4}$, $x_{i5}$, and $x_{i6}$ are defined as before, but now

$$x_{i1} = \begin{cases} 1 & \text{if series 1} \\ 0 & \text{otherwise} \end{cases}$$

and in addition

$$x_{i7} = \begin{cases} 1 & \text{if series 2} \\ 0 & \text{otherwise} \end{cases}$$

$$x_{i8} = \begin{cases} 1 & \text{if series 3} \\ 0 & \text{otherwise} \end{cases}$$

From a purely statistical point of view one may combine various aspects of the Hansch and Free-Wilson models into a single equation. This is simply equivalent to the simultaneous appearance of both continuous and binary variables in the model. Also, from a purely statistical point of view one may apply regression techniques to compounds that do not form a series of congeners. In both cases, the usefulness of such combinations will ultimately depend on the degree to which the resulting equations help predict active compounds. To a large extent this is an empirical question.

### 1.2.1.2. Coefficient Estimation

We now assume that a model of the following form has been specified:

$$Y_1 = \beta_1 x_{11} + \beta_2 x_{12} + \cdots + \beta_m x_{1m} + \varepsilon_1$$
$$Y_2 = \beta_1 x_{21} + \beta_2 x_{22} + \cdots + \beta_m x_{2m} + \varepsilon_2$$
$$\vdots$$
$$Y_n = \beta_1 x_{n1} + \beta_2 x_{n2} + \cdots + \beta_m x_{nm} + \varepsilon_n$$

Letting Y denote the $n \times 1$ column vector $(Y_1, \ldots, Y_n)^T$, $\beta$ the $m \times 1$ column vector $(\beta_1, \ldots, \beta_m)^T$, X the $n \times m$ matrix

$$
\begin{bmatrix}
x_{11} & x_{12} & \cdots & x_{1m} \\
x_{21} & x_{22} & \cdots & x_{2m} \\
\cdots & \cdots & \cdots & \cdots \\
x_{n1} & x_{n2} & \cdots & x_{nm}
\end{bmatrix}
$$

and $\varepsilon$ the $n \times 1$ column vector $(\varepsilon_1, \varepsilon_2, \ldots, \varepsilon_n)^T$, the set of equations above may be written

$$Y = X\beta + \varepsilon$$

In this context X is called the design matrix for the regression.

Our objective is to estimate the unknown coefficient vector $\beta$. The least-squares estimator $\beta^*$ for $\beta$ is that vector $\gamma$ which minimizes the residual sum of squares

$$Q(\gamma) = \sum_{i=1}^{n} (Y_i - \gamma_1 x_{i1} - \gamma_2 x_{i2} - \cdots - \gamma_m x_{im})^2$$

In matrix form, the above may be written

$$Q(\gamma) = (Y - X\gamma)^T (Y - X\gamma)$$

where T denotes transpose. Differentiating $Q(\gamma)$ with respect to $\gamma_j$ and equating the resulting expression to 0 $(j = 1, \ldots, m)$, one obtains m equations which may be written in matrix form as

$$X^T X \gamma = X^T Y$$

These are the "normal equations" for the least-squares estimators, and any solution of the equations above is a set of coefficients that minimizes $Q(\gamma)$. Henceforth, any such solution will be denoted $\beta^*$.

These coefficients have the following geometrical interpretation. Let each of the random variables $Y_1, \ldots, Y_n$ define a coordinate in $R^n$ (i.e., n-dimensional Euclidean space). We assume that the expectation of the joint distribution of Y may be expressed as $\theta = X\beta$, and therefore lies in the subspace S of $R^n$

spanned by the columns of X. Furthermore, given an observation Y, $\theta$ is estimated by $\theta^* = X\beta^*$. Since $X^T(Y - \theta^*) = X^TY - X^TX\beta^* = 0$, then $Y - \theta^*$ is orthogonal to every vector of the form $X\gamma$, and hence $Y - \theta^*$ is orthogonal to S. Thus we estimate $\theta^*$ by the projection of Y on S or by the point of S nearest to Y. This projection determines $\theta^*$ uniquely. We may therefore define the least-squares estimators to be any set of coefficients $\beta^*$ which yields $X\beta^* = \theta^*$, where $\theta^*$ is the orthogonal projection of Y onto S, the space spanned by the columns of X.

To actually estimate $\beta^*$, we shall need to find the solution of the normal equations. Two cases arise.

First is the case where $X^TX$ is a nonsingular matrix with unique inverse $(X^TX)^{-1}$. Premultiplying the equation $X^TX\beta = X^TY$ by $(X^TX)^{-1}$ we obtain the unique solution

$$\beta^* = (X^TX)^{-1}X^TY$$

One may show that $\beta^*$ is an unbiased estimate of $\beta$, and each component $\beta_j^*$ has the minimum variance of any linear estimate of $\beta_j$. This case is typical of equations, such as Hansch's, in which the $x_{ij}$'s may vary continuously.

The second case is more complex. Here we assume that $X^TX$ is a singular matrix and hence does not possess a unique inverse. This implies that the columns of X are linearly dependent, and thus while $\theta^*$ is always unique there is more than one $\beta^*$ solution to $X\beta^* = \theta^*$, or, equivalently, more than one $\beta^*$ solution to $X^TX\beta^* = X^TY$. As an example where this arises in practice we may consider the Free-Wilson model. Let us assume that the observed compounds have the following structure:

| Compound | A site | B site |
|----------|--------|--------|
| 1 | $a_1$ | $b_1$ |
| 2 | $a_1$ | $b_2$ |
| 3 | $a_2$ | $b_1$ |
| 4 | $a_3$ | $b_1$ |
| 5 | $a_3$ | $b_2$ |

This yields the design matrix

| v | $a_1$ | $a_2$ | $a_3$ | $b_1$ | $b_2$ |
|---|---|---|---|---|---|
| 1 | 1 | 0 | 0 | 1 | 0 |
| 1 | 1 | 0 | 0 | 0 | 1 |
| 1 | 0 | 1 | 0 | 1 | 0 |
| 1 | 0 | 0 | 1 | 0 | 1 |
| 1 | 0 | 0 | 1 | 0 | 1 |

where v is a dummy variable used to denote the presence of the overall mean. Letting an underscored variable denote a column vector, we note, for example, that $\underline{b}_1 + \underline{b}_2 = \underline{v}$, and hence the columns of the matrix above are not independent. Similarly, $\underline{a}_1 + \underline{a}_2 + \underline{a}_3 = \underline{v}$, since the $a_i$'s are mutually exclusive and one of these must occur at site A for each compound.

Let us now suppose that $\theta^*$ is given and $\beta^*$ is a solution to $X\beta^* = \theta^*$, where X is the foregoing design matrix. Since the columns of X are not independent, there exists $\gamma \neq 0$ such that $X\gamma = 0$. Therefore, $\hat{\beta} = \beta^* + \gamma \neq \beta^*$ satisfies $X\hat{\beta} = X(\beta^* + \gamma) = X\beta^* + X\gamma = \theta^* + 0$ as well. For example, set $\gamma^T = (2, -1, \ldots, -1)$. What is required to avoid these indeterminancies is to uniquely choose a specific solution by placing appropriate restrictions on the set of possible solutions. A convenient one in the Free-Wilson setting is to impose the restrictions $\Sigma_{i=2}^{4} \beta_i = 0$ and $\Sigma_{i=5}^{6} \beta_i = 0$, for then the observed coefficients will describe the relative merits with respect to zero of the alternatives at substituent sites A and B.

In general, any set of linear constraints can be expressed as $H\beta = 0$, where H is a p × m matrix containing p linearly independent rows. In the example above,

$$H = \begin{bmatrix} 0 & 1 & 1 & 1 & 0 & 0 \\ 0 & 0 & 0 & 0 & 1 & 1 \end{bmatrix}$$

This constraint may be incorporated into the model in two different ways. First, one may reparameterize the model by setting $\hat{\beta}^T = (\beta_1, \beta_2, \beta_3, -\beta_2 - \beta_3, \beta_5, -\beta_5)$, where we use $\beta_4 = -\beta_2 - \beta_3$ and $\beta_6 = -\beta_5$. Letting $\theta^*$ be unknown but fixed, we see that the left-hand side of $X\hat{\beta} = \theta^*$ may be written as

$$\sum_{i=1}^{6} x_{ij}\hat{\beta}_j = x_{i1}\beta_1 + (x_{i2} - x_{i4})\beta_2 + (x_{i3} - x_{i4})\beta_3$$

$$+ (x_{i5} - x_{i6})\beta_5$$

$(i = 1, 2, \ldots, n)$. Setting $z_{i1} = x_{i1}$, $z_{i2} = (x_{i2} - x_{i4})$, $z_{i3} = (x_{i3} - x_{i4})$, $z_{i4} = (x_{i5} - x_{i6})$, $\gamma_1 = \beta_1$, $\gamma_2 = \beta_2$, $\gamma_3 = \beta_3$, and $\gamma_4 = \beta_5$, we obtain the equivalent system of equations

$$Z\gamma = \theta*$$

with the reduced design matrix

$$\begin{bmatrix} 1 & 1 & 0 & 1 \\ 1 & 1 & 0 & -1 \\ 1 & 0 & 1 & 1 \\ 1 & -1 & -1 & 1 \\ 1 & -1 & -1 & -1 \end{bmatrix}$$

This matrix is of full rank (i.e., all column vectors are independent), and the solution is $\gamma* = (Z^T Z)^{-1} Z^T Y$.

An alternative approach to reparameterization is to redefine the normal equations to explicitly incorporate the restriction $H\beta = 0$. We introduce the $p \times 1$ Lagrangian vector $\lambda = (\lambda_1, \lambda_2, \ldots, \lambda_p)^T$, and by minimizing $Q(\gamma)$ subject to the constraint $H\beta = 0$, obtain the expanded normal equations

$$X^T X \beta* + H^T \lambda* = X^T Y$$

$$H\beta* = 0$$

Since $H^T(H\beta*) = 0$, we may also write

$$X^T X \beta* + H^T H \beta* + H^T \lambda* = X^T Y$$

$$H\beta* = 0$$

or

$$\begin{bmatrix} A^T A & H \\ H & 0 \end{bmatrix} \begin{bmatrix} \beta* \\ \lambda* \end{bmatrix} = \begin{bmatrix} X^T Y \\ 0 \end{bmatrix}$$

where

$$A = \begin{bmatrix} X \\ H \end{bmatrix}$$

If H contains sufficient restrictions to identify $\beta*$ uniquely, then the matrix A is of full rank, and it may be proved that the matrix

$$\begin{bmatrix} A^T A & H \\ H & 0 \end{bmatrix}$$

is nonsingular and may be inverted. Hence

$$\begin{bmatrix} \beta* \\ \lambda* \end{bmatrix} = \begin{bmatrix} A^T A & H^T \\ H & 0 \end{bmatrix}^{-1} \begin{bmatrix} X^T Y \\ 0 \end{bmatrix}$$

and the estimators $\beta*$ are determined. The difficulty with this procedure is, of course, that regular regression packages cannot be employed, and an enlarged rather than a reduced matrix must be inverted.

### 1.2.1.3. Estimation of $\sigma^2$ and the Variance of the Least-Squares Estimators

Henceforth, let us assume that the design matrix, after a possible reparameterization, is of full rank. Then $(X^T X)^{-1}$ exists and $\beta*$ is the unique solution of $\beta* = (X^T X)^{-1} X^T Y$. From this it immediately follows that the covariance matrix of $\beta*$ is

$$(X^T X)^{-1} X^T \, \text{Cov}(Y) X (X^T X)^{-1} = (X^T X)^{-1} X^T \sigma^2 I X (X^T X)^{-1}$$

$$= \sigma^2 (X^T X)^{-1}$$

where I is the n × n identity matrix.

In the above, the variance $\sigma^2$ is unknown but an estimator for it may be obtained. To do so, note that the expectation of $Q(\beta)$ [i.e., $E(Y - X\beta)^T (Y - X\beta) = \epsilon^T \epsilon$] is $n\sigma^2$. Now $Y - X\beta$ is the sum of the orthogonal components $Y - X\beta*$ and $X(\beta* - \beta)$. It follows that

$$Q(\beta) = Q_0 + Q_1(\beta)$$

where

$$Q_0 = (Y - X\beta^*)^T(Y - X\beta^*)$$

$$Q_1(\beta) = (\beta - \beta^*)X^TX(\beta - \beta^*)$$

Of these, $Q_0$ does not contain the unknown $\beta$ and it may be shown that $EQ_0 = (n - m)\sigma^2$. Hence $Q_0/(n - m)$ is an unbiased estimator of $\sigma^2$.

### 1.2.1.4. Hypothesis Testing

After fitting the regression, one will typically want to test a variety of hypotheses concerning the model. For example, one may wish to test whether or not variable $x_j$ contributes to the determination of Y. This is equivalent to testing the hypothesis $\beta_j = 0$. A test statistic may be constructed by comparing the reduction in the variance obtained by permitting $\beta_j$ to be any value with that obtained when we impose the restriction $\beta_j = 0$. Specifically, for $Q(\gamma) = (Y - X\gamma)^T(Y - X\gamma)$, set $R_0 = \min_\gamma Q(\gamma)$ and set $R_1 = \min_\gamma Q(\gamma)$ under the constraint $\beta_j = 0$. It may be shown that under the hypothesis $\beta_j = 0$, the quantities $R_1 - R_0$ and $R_0$ are independent with respective means $\sigma^2$ and $(n - m)\sigma^2$. The ratio

$$F = \frac{(R_1 - R_0)/1}{R_0/(n - m)}$$

therefore has a mean value which is close to 1. On the other hand, if the null hypothesis is false, $R_1$ will be substantially larger than $R_0$, reflecting a bad fit, and the ratio F will be large. Hence the null hypothesis should be rejected if F is substantially larger than 1.

More complex linear hypotheses may also be tested by the method above. Let the null hypothesis be of the form

$$H_0: A^T\beta = b$$

where A is a m × p matrix of full rank (i.e., of rank p, as $p \leqslant m$) and b is a m × 1 vector. Set $R_0 = \min_\gamma Q(\gamma)$ and set $R_1 = \min_\gamma Q(\gamma)$ under the constraint $A^T\beta = b$. Then under the null hypothesis, the quantities $R_1 - R_0$ and $R_0$ are independent with means $p\sigma^2$ and $(n - m)\sigma^2$. The results of these calculations are often laid out in an "analysis of variance" table as follows:

| | Sum of squares | Degrees of freedom | Mean square | Ratio |
|---|---|---|---|---|
| Difference | $R_1 - R_0$ | p | $\dfrac{R_1 - R_0}{p}$ | $F = \dfrac{n - m}{p} \cdot \dfrac{R_1 - R_0}{R_0}$ |
| Residual | $R_0$ | n − m | $\dfrac{R_0}{n - m}$ | |
| Total | $R_1$ | n − m + p | | |

Again the ratio F has a mean of around 1 under the null hypothesis, and the null hypothesis is rejected when F is large.

In the above, we assumed only that the error terms are uncorrelated with mean 0 and common unknown variance $\sigma^2$. When we ask how large F should be for rejection we need more assumptions. In particular, it is convenient to assume that the error terms are normally distributed. In that case it may be shown that F has an F distribution with p and n -- m degrees of freedom. Hence the null hypothesis is rejected at the $\alpha$ level whenever $F > F_\alpha(p, n - m)$.

To carry out the calculations we shall need to determine $R_0$ and $R_1 - R_0$. To do so, first compute the least-squares estimator $\beta^*$. Then

$$R_0 = (Y - X\beta^*)^T(Y - X\beta^*)$$

and

$$R_1 - R_0 = (A^T\beta^* - b)^T[A^T(X^TX)^{-1}A]^{-1}(A^T\beta^* - b)$$

Two examples will illustrate the technique. Consider the hypothesis $\beta_j = 0$. Set $A = (0, 0, \ldots, 1, 0, \ldots, 0)^T$ with 1 appearing in the jth position. After simplification one finds

$$F = \frac{(\beta_j^*)^2}{S_{jj}(Y - X\beta^*)^T(Y - X\beta^*)}$$

where $S_{jj}$ is the $(j,j)$th element of the matrix $(X^TX)^{-1}$.

Now consider the hypothesis that none of the independent variables helps estimate Y (i.e., $\beta_2 = \beta_3 = \cdots = \beta_m = 0$). This may be written as $A^T\beta = \underline{0}$, where

$$A = \begin{bmatrix} 0 & 0 & \cdots & 0 \\ 1 & 0 & \cdots & 0 \\ 0 & 1 & \cdots & 0 \\ 0 & 0 & \cdots & 1 \end{bmatrix}$$

Then

$$R_0 = (Y - X\beta^*)^T(Y - X\beta^*) \quad \text{and} \quad R_1 - R_0 = (Y - \bar{Y})^T(Y - \bar{Y})$$
$$- (Y - X\beta^*)^T(Y - X\beta^*)$$

which gives

$$F = \frac{(Y - \bar{Y})^T(Y - \bar{Y}) - (Y - X\beta^*)^T(Y - X\beta^*)}{(Y - X\beta^*)^T(Y - X\beta^*)} \cdot \frac{n - m}{m - 1}$$

This is the ordinary test for the significance of the multiple correlation coefficient

$$R^2 = 1 - \frac{(Y - X\beta^*)^T(Y - X\beta^*)}{(Y - \bar{Y})^T(Y - \bar{Y})}$$

### 1.2.1.5. Goodness of Fit

It is frequently pointed out that a measure of the goodness of fit of the estimated regression is given by $R^2$, which was introduced at the end of the preceding section as the ratio of the sum of squares due to the regression to the total sum of squares about the mean. Clearly, $R^2$ takes a value between 0 and 1, where 0 means that there is no reduction in the sum of squares when the $Y_i$ values are predicted by the regression line rather than the overall mean, and where 1 means that the $Y_i$ values are perfectly predicted by the regression line. Everything else being equal, a high $R^2$ value is to be preferred to a low $R^2$ value.

Also introduced in the preceding section was the accompanying test for testing whether or not $R^2$ is significantly different from zero. As remarked there, this is equivalent to testing whether or not $\beta_2 = \beta_3 = \cdots = \beta_m = 0$. It should be noted that obtaining a significant result is not equivalent to validating the model. For example, testing $R^2 = 0$ in the Free-Wilson setting and obtaining a significant F-ratio does not imply that the assumption of additivity characterizing the Free-Wilson model is correct. This is so because the test $R^2 = 0$ is performed *under the assumption that the model is correct*. In fact, all tests discussed in the preceding section are performed under such an assumption.

The above does not imply that we find ourselves in a closed circle without any statistical method of verifying the model. To understand the situation, let us consider the nature of the errors that can occur. A set of observations may not satisfy the relationship

$$y_i = \sum_{j=1}^{m} \beta_j x_{ij}$$

with precision for two very different reasons. First, $y_i$ may be observed with error. Second, $y_i$ may be observed with precision but not be linearly related to the $x_{ij}$ values as assumed. Let us describe the magnitude of these two discrepancies by $\varepsilon_i$ and $\xi_i$, respectively. Hence we have

$$Y_i = \sum_{j=1}^{m} \beta_j x_{ij} + \varepsilon_i + \xi_i$$

In obtaining the best fit we estimate $\Sigma(\varepsilon_i + \xi_i)^2$, but we cannot tell whether the error is due primarily to $\varepsilon$ or to $\xi$. For example, one may obtain an $R^2$ value that is low due primarily to $\varepsilon$ (e.g., high biological variation). Thus the model is correct (i.e., $\xi = 0$), but still the $R^2$ is low. Another possibility is that $R^2$ is high and that the reason it is not equal to 1 is due solely to the model being wrong (i.e., $\varepsilon = 0$ and $\xi > 0$). For example, suppose that $y_i = \ln x_i$, $\varepsilon_i = 0$ and we try to fit the model $y_i = \beta_1 + \beta_2 x_i$ for values $x_i = 3, 4, 5, 6, 7, 8, 9, 10, 11$. Then $\beta_1^* = 0.76994$, $\beta_2^* = 0.15682$, $\sigma^{*2} = 0.006232$, and $R^2 = 0.9970339$. Here $R^2$ has an F-ratio of 336.14543, which has a significance level smaller than 0.0005. But the model is wrong.

One method of detecting that the model is wrong is to examine the residuals to detect unusual patterns. For the example above, we find the sequence ++------++ in the signs of the residuals. A close examination of the residuals should always be carried out, and most regression programs will plot them.

An alternative method is to obtain a measure of $\epsilon$ or $\xi$ which is independent of the model being fitted. Knowledge of $\xi$ requires knowledge of the true model, which we do not have. However, an independent estimate of $\epsilon$ can be obtained by taking repeated observations. This we can do. A test can then be constructed by comparing the true variance, as estimated by the repeated measurements, with the residual variance obtained under the assumption that the model is true. If the model is correct, they should have similar values. Therefore, let $Y_{ij}$ be the jth observation on compound i and let

$$\overline{Y}_i = \frac{1}{n_i} \sum_{j=1}^{n_i} Y_{ij}$$

where $n_i$ is the number of observations on the ith compound. Hence

$$R_0 = \sum_{j=1}^{n} n_i (Y_{ij} - \overline{Y}_i)^2$$

has expectation $\sigma^2 \sum_{i=1}^{n} (n_i - 1)$. Let $R_1 = \min_\beta Q(\beta)$ be the residual sum of squares under the restriction that the model is correct. Then a test statistic is

$$\frac{(R_1 - R_0)/(n - m)}{R_0 / \sum_{i=1}^{n} (n_i - 1)}$$

which has an F distribution with $n - m$ and $\sum_{i=1}^{n} (n_i - 1)$ degrees of freedom, and the null hypothesis that the model is correct is rejected if this is large.

Even though a model may be found to be incorrect, it may still be very useful in predicting active compounds. However, knowledge that it is incorrect will alert us to the fact that there may be important factors not yet considered. On the other hand, if the residuals are well behaved and the null hypothesis is not rejected, the evidence is that the relative magnitude of errors is such that no alternative model can be found that will yield a significantly better fit.

### 1.2.2. Variable Selection

In the preceding section we discussed the question of a single fixed equation. A related question is what variables should be included in a regression. Often one takes may different measurements on a compound, several of which are similar. It is well known that with a sufficiently large number of independent variables and a fixed number of dependent observations, any desired degree of fit can be obtained.

A guiding principle is therefore that of parsimony, explaining an effect with the smallest number of variables possible. But which variables should be chosen? In this section we discuss three different methods.

### 1.2.2.1. All-Subsets Regression

The conceptually simplest method for discovering a good subset regression equation is to perform a regression on all possible subsets. Then several different types of criteria could be used for selecting one of these subsets. For example, among the subsets containing j independent variables one would select the best in terms of, say, yielding the highest $R^2$ for a regression using j independent variables. Denoting these resulting pairs by $(j, R_j^2)$, one would plot these values and select as the best regression that pair $(j, R_j^2)$ where a "bend" occurs in the graph (see Figure 1). An alternative criterion would be obtained by plotting $(C, R^2)$, where C is the cost of obtaining information on the independent variables in the regression. Then some cutoff value would be chosen.

One of the severe drawbacks with the all-subsets approach is that the computational effort required is prohibitively large for even a moderately large number of independent variables. If m is the number of such variables, there are $2^m$ subsets. Thus for m = 10 we obtain 1024 different regressions, while for m = 15 we have 32,768 different regressions.

Beale et al. (1967) have written a program that will find the pairs $(j, R_j^2)$ in a much more efficient fashion. By the use of cutoff rules many suboptimal subsets that cannot be optimal according to the $R_j^2$ criterion are eliminated at an early stage of the

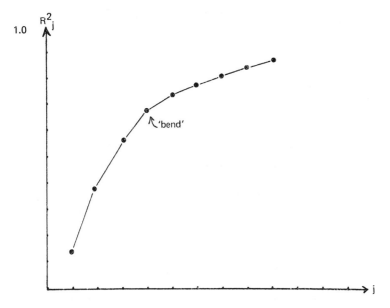

**Figure 1** Selecting the best subset.

computational effort. However, this deprives the user of the possibility of considering other criteria, such as cost, in selecting among the subsets.

### 1.2.2.2. Stepwise Regression

An alternative to the best subset method is stepwise regression. Versions of it are available in many different statistical computer packages.

The most widely used version is one by Efroymson (1960) in which both backward and forward steps are taken. At each stage of the procedure a single regressor is either added to or eliminated from the current regression model in accordance with the following rules. Two F levels, say, $F_{in}$ and $F_{out}$, are initially specified by the user. Then a regressor, say $X_j$, is eliminated from the equation if its F ratio for testing $\beta_j$ in the current regression model is less than or equal to $F_{out}$. If there is a choice of variables to be eliminated, the one with the smallest F ratio is chosen. If no variable can be eliminated in this way, then a regressor, say $X_j$, is brought in if its F ratio for testing $\beta_j = 0$ in the $X_j$-augmented current regression equation is greater than or equal to

$F_{in}$. Again, if there is a choice of variable to be introduced, the one giving the largest F-ratio is chosen. The process ends when a regressor cannot be brought into the model.

It should be noted that this method does not guarantee that the final equation consisting of j variables will be the subset of j variables that gives the highest $R_j^2$. It should further be noted that the user does not receive a collection of good subsets to choose from, only a single final subset.

### 1.2.2.3.  Spurious Significance with Stepwise Regression

Since a "good subset" of variables is selected through the use of a stepwise regression procedure, there is an increased chance of obtaining a spuriously significant regression equation. Topliss and Edwards (1979) have investigated this problem. Their method was to generate a set of observations, $y_1, \ldots, y_n$ and $x_{11}, \ldots, x_{1m}, x_{21}, \ldots, x_{2m}, x_{n1}, \ldots, x_{nm}$, each of which is independent and uniformly distributed, and to perform a stepwise regression on them. Their results are summarized in Figures 2 through 4.

Figure 2 shows the relationship between the number of observations and number of independent variables for a 1% probability of $R^2$ attaining various different values. This graph permits the determination of the number of observations required to screen, for example, 10 variables while keeping the probability of

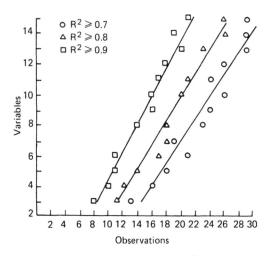

**Figure 2** 99th percentile of $R^2$. (Figures 2, 3, and 4 are reprinted with permission from *J. Med. Chem.* (1979), Vol. 22, p. 1238. ©1979 American Chemical Society.)

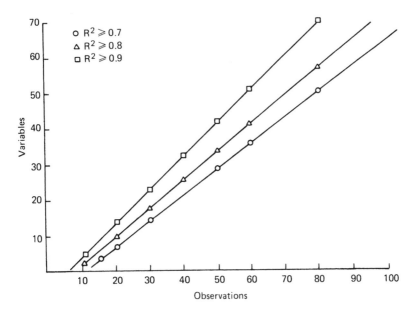

**Figure 3** 99th percentile of $R^2$.

encountering a chance correlation with $R^2 \geqslant 0.8$ at the 1% level or less. Figure 3 is an extrapolation of the previous graph for larger numbers of screened variables. For different levels of tolerated chance correlation, Figure 4 is useful. It shows how the number of required observations changes as the tolerated chance correlation level is allowed to increase from 1% to 5%, and then to 10%.

A different approach to the spurious $R^2$ problem in stepwise regression is the following. First, randomly split the observations in half and run the stepwise procedure on the first half to identify an appropriate variable subset. Then verify the fit using the second half either by running a standard regression using these variables or by observing the fit of the predicted values using the regression equation obtained initially. A test for the equation given by the first of these methods is given by the unadjusted $R^2$ and the associated F-ratio. The significance of fit using the second method may be tested by the test statistic

$$F = \frac{n - m}{m} \cdot \frac{R_1 - R_0}{R_0}$$

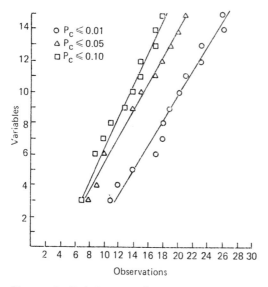

Figure 4 Points at which 0.8 is the 99th, 95th or 90th percentile for $R^2$.

where

$$R_0 = Q(\beta^{(2)}) = \sum_{i=1}^{n_2} \left[ Y_i^{(2)} - \sum_{j=1}^{m} \beta_j^{(2)} x_{ij}^{(2)} \right]^2$$

$$R_1 = Q(\beta^{(1)}) = \sum_{i=1}^{n_2} \left[ Y_i^{(2)} - \sum_{j=1}^{m} \beta_j^{(1)} x_{ij}^{(2)} \right]^2$$

and where: $\beta^{(1)}$ and $\beta^{(2)}$ are the least-squares estimators obtained in fitting the first half and second half of the data set, respectively; $n_2$ is the number of observations in the second half of the data set; and $Y_i^{(2)}$ and $x_{ij}^{(2)}$ are observations belonging to the second half of the data set. This test statistic has an F distribution with m and n − m degrees of freedom.

## 1.2.2.4. Principal Components

Hawkins (1973) has developed a different approach to finding a "best subset" regression equation. It relies on the following geometrical interpretation of the regression problem. Let $Y$, $X_1, \ldots, X_m$ be the standardized variables obtained from the original set of variables by subtracting from each its sample mean and dividing the result by the corresponding sample standard deviation. Now associate each of these variables with one of the coordinate axes in $R^{m+1}$. In this context the problem of finding the least-squares estimates is equivalent to finding a hyperplane

$$Y - \beta_1 X_1 - \beta_2 X_2 - \cdots - \beta_m X_m = 0$$

passing through the origin in $R^{m+1}$ such that the sum of squares of the deviations along the Y axis is minimized (see Figure 5); that is, it minimizes

$$Q(\beta) = \sum_{i=1}^{n} (Y_i - \beta_1 X_{i1} - \beta_2 X_{i2} - \cdots - \beta_m X_{im})^2$$

In finding the collection of best subregressions, we are trying to find hyperplanes passing through the origin which lie close to the $Y_i$ values in terms of the vertical norm defined above. The set of coefficients of such alternative hyperplanes which are good

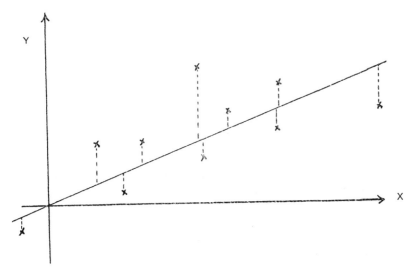

**Figure 5** The vertical norm. (Figures 5 and 6 are courtesy of the Royal Statistical Society, England.)

fits defines a set S or $R^{m+1}$. The problem Hawkins has considered is how one might characterize this set. If this problem is solved satisfactorily, one obtains a collection of good subset regressions, and one is in a position to choose one of these regressions on the basis of some suitable criterion that does not depend on fit. For example, if the two variable subsets $(X_1, X_2, X_6)$ and $(X_3, X_5, X_6)$ when regressed yield comparable fits, one may decide to choose the former over the latter on the basis that measurements on $X_1$ and $X_2$ are cheaper than measurements on $X_3$ and $X_5$.

Hawkins points out that there is no known satisfactory characterization of these hyperplanes for the present regression problem. However, one may define a slightly different regression problem for which such a characterization exists. Hopefully, these two regression problems are sufficiently similar in practice that the close-fitting set S' formed for the latter is similar to the close-fitting set S for the former. If that were true, selecting a good regression equation in S' would yield a good regression equation belonging to S as well.

This slightly different regression problem may be defined as follows. In $R^{m+1}$ find the hyperplane

$$Y - \beta_1 X_1 - \cdots - \beta_m X_m = 0$$

such that the sum of the orthogonal distances of the data points $(Y_i, X_{i1}, \ldots, X_{im})$, $i = 1, \ldots, n$, from this hyperplane is minimized; that is, find the vector $\beta$ that minimizes

$$\lambda(\beta) = \sum_{i=1}^{n} \frac{(Y_i - \beta_1 X_{i1} - \cdots - \beta_m X_{im})^2}{1 + \beta_1^2 + \cdots + \beta_m^2}$$

This orthogonal norm is easily displayed in two dimensions (see Figure 6).

The solution to the foregoing problem is well known. First, let Z denote the $n \times (m + 1)$ matrix of observed values for the vectors $(Y_i, X_{i1}, \ldots, X_{im})$, $i = 1, \ldots, n$. Hence we seek $\gamma = (1, \gamma_2, \ldots, \gamma_{m+1})^T$ such that

$$\frac{(Z\gamma)^T (Z\gamma)}{\gamma^T \gamma}$$

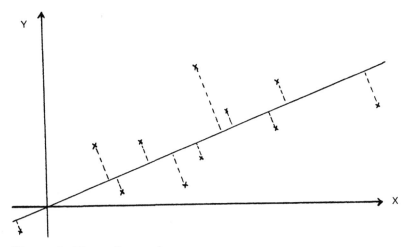

**Figure 6** The orthogonal norm.

is minimized. Note, however, that for $\gamma$ with arbitrary $\gamma_1 \neq 0$, we have

$$\frac{(Z\gamma)^T(Z\gamma)}{\gamma^T\gamma} = \frac{[Z(\gamma/\gamma_1)]^T[Z(\gamma/\gamma_1)]}{(\gamma/\gamma_1)^T(\gamma/\gamma_1)}$$

Hence we may drop the restriction $\gamma_1 = 1$. Furthermore, suppose that $\gamma^T\gamma = C$, some positive constant. Then also

$$\frac{[Z(\gamma/C^{1/2})]^T[Z(\gamma/C^{1/2})]}{(\gamma/C^{1/2})^T(\gamma/C^{1/2})} = \frac{(Z\gamma)^T(Z\gamma)}{\gamma^T\gamma}$$

Hence we may assume without loss of generality that $\gamma^T\gamma = 1$. To minimize $\gamma^T Z^T Z\gamma/\gamma^T\gamma$ subject to the constraint $\gamma^T\gamma = 1$, we may use Lagrange multipliers. Set $\psi = \gamma^T R\gamma - \mu(\gamma^T\gamma - 1)$, where $R = Z^T Z$, the correlation matrix of the observed values. The vector of partial derivatives is

$$\frac{\partial \psi}{\partial \gamma} = 2R\gamma - 2\mu\gamma$$

which, upon being set equal to zero, reduces to

$(R - \mu I)\gamma = 0$

Letting $\mu^*$ and $\gamma^*$ denote the required solution, we see that $\mu^*$ is an eigenvalue of R and $\gamma^*$ the associated eigenvector. To determine which of the eigenvalues $\lambda_1 \geqslant \cdots \geqslant \lambda_{m+1}$ is $\mu^*$, premultiply $(R - \mu^*I)\gamma^* = 0$ by $\gamma^{*T}$, and use the fact that $\gamma^{*T}\gamma^* = 1$, to obtain $\gamma^{*T}(R - \mu^*I)\gamma^* = \gamma^{*T}R\gamma^* - \mu^* = 0$, or $\gamma^{*T}R\gamma^* = \mu^*$. But $\gamma^{*T}R\gamma^*$ is what is to be minimized and hence $\mu^* = \lambda_{m+1}$, the smallest eigenvalue of R. Thus $\min_\gamma \lambda(\gamma) = \lambda(\gamma_{m+1}) = \lambda_{m+1}$, where $\gamma_i$ is the eigenvector corresponding to $\lambda_i$ $(i = 1, 2, \ldots, m + 1)$.

To find the set of hyperplanes (or equivalently, the set of vectors defining the coefficients of these hyperplanes) which are close to this optimal fit, we go further and find another vector which minimizes the orthogonal distance from the observed values in $R^{m+1}$ to the hyperplane defined by this vector under the restriction that this vector is orthogonal to the first just obtained. It may be shown that this is the eigenvector corresponding to the second smallest eigenvalue. Continuing in this way, a basis for the entire space $R^{m+1}$ may be constructed. Furthermore, the hyperplanes defined by the subspace spanned by the p ($<$m) eigenvectors associated with p eigenvalues which are relatively small all lie close to the data. In particular, consider the hyperplane defined by the vector of coefficients $\sum_{i=m-p+1}^{m+1} c_i\gamma_i$, with $\sum_{i=m-p+1}^{m+1} c_i^2 = 1$. The orthogonal norm for the associated hyperplane is $\lambda = \sum_{i=m-p+1}^{m+1} c_i^2\lambda_i$, and if each of the $\lambda_i$ is small, so is $\lambda$.

The relationship between the orthogonal norm and the vertical norm may be described as follows. Let $\theta(\beta)$ be the angle between the Y axis and the perpendicular to a given hyperplane $Y - \beta_1 X_1 - \cdots - \beta_m X_m$. Then

$$\lambda(\beta) = Q(\beta)\cos^2\theta(\beta)$$

This is less than $Q(\beta)$. Thus if $Q(\beta)$ is small, $\lambda(\beta)$ is small. Consequently, a hyperplane H with a small vertical norm would also have a small orthogonal norm.

Expressing H as $\sum_{i=1}^{m+1} c_i\gamma_i$, its orthogonal norm is $\sum_{i=1}^{m+1} c_i^2\lambda_i$, and since this is small, $c_i$ is small whenever $\lambda_i$ is large. Thus the hyperplane H loads predominantly on the eigenvectors corresponding to the low eigenvalues, and therefore is almost embedded in the space spanned by the eigenvectors corresponding to these low eigenvalues. That is, we will find a good approximation to H in this space.

The virtue of the foregoing relationship is that the subspace
S' may be explored in a systematic fashion to discover good alternative subset regression equations. To do this, define the
$(m + 1) \times (m + 1)$ matrix D by

$$d_{ij} = \frac{\gamma_{ij}}{\sqrt{\lambda_i}}$$

where $\gamma_{ij}$ is the jth component of the ith eigenvector of the matrix
R. Hawkins has shown that the residual variance in the vertical
norm of the best-fitting equation is $1/\sum_{i=1}^{m+1} d_{i1}^2$. Furthermore, if

the vector $(d_{i2}^*,\ldots,d_{2,m+1}^*)$ obtained from the ith row of D by set-

ting $d_{ij}^* = d_{ij}/d_{i1}$ is used to define the coefficient of a regression

equation, then $1/d_{i1}^2$ will be the residual vertical sum of squares

of that equation. If some of the loadings in this row are zero,
the best regression on the remaining variables will have a residual vertical sum of squares no greater than $1/d_{i1}^2$.

The results above were stated for the matrix D derived from
the eigenvectors of R. However, the mathematics in obtaining
them assumed only that the rows were orthonormal. Hence these
results apply to any orthonormal rotation of D as well. Therefore, to identify good subregressions it would be desirable to improve D by obtaining as many zeros as possible by means of a rotation. This may be done by the varimax rotation procedure used
in factor analysis.

After rotation, giving a new D, most of the $d_{i1}$ $(i = 1, 2, \ldots, m
+ 1)$ will have values close to zero. If one $d_{i1}$ is large and domin-
ates the rest, then $1/d_{i1}^2$ is not much larger than $1/\sum d_{i1}^2$, and the
regression equation based on the ith row is essentially a unique
good predictor. Those variables with a high loading in this row
will form a subset with good predictive qualities. Other subsets
may be obtained by an additional orthogonal transformation of D
that corresponds to a substitution process. In particular, if $d_{ij}$
is large, alternative variables to the jth may be found by locating
a row k such that $d_{k1}$ is small and $d_{kj}$ is large. The matrix
$T = \{t_{ij}\}$ is derived from the identity matrix by making the following alterations:

$$t_{ii} = \frac{1}{(1 + r^2)^{1/2}}$$

$$t_{ik} = \frac{-r}{(1 + r^2)^{1/2}}$$

$$t_{kj} = \frac{r}{(1 + r^2)^{1/2}}$$

$$t_{kk} = \frac{1}{(1 + r^2)^{1/2}}$$

where $r = d_{ij}/d_{kj}$. The matrix $\hat{D} = TD$ is orthonormal with $\hat{d}_{ij} = 0$ and

$$\hat{d}_{i1} = \frac{d_{i1} - rd_{k1}}{(1 + r^2)^{1/2}}$$

We note that $1/\Sigma\ d_i^2 = 1/\Sigma\ \hat{d}_i$, and if r is small, $d_{i1} \cong \hat{d}_{i1}$. Thus row i of $\hat{D}$ is also a good predictor, but with different variables taking a high loading in place of $X_j$.

An example by Hawkins will illustrate the technique. [For additional examples, see Jeffers (1981).] It is required to predict $X_1$, the yield of a washing plant, from input ($X_2$ to $X_6$) and output ($X_7$ to $X_{10}$) characteristics of the feedstock for a mine. There are known near multicollinearities between $X_4$, $X_5$, and $X_6$, and between $X_8$, $X_9$, and $X_{10}$. There is also a known high correlation between $X_3$ and $X_4$ and between $X_7$ and $X_8$. The cost of measurement of the predictors is as follows:

Very cheap:    $X_2$, $X_4$, and $X_8$

Fairly cheap:  $X_5$ and $X_9$

Expensive:    the remainder

and the predictor selected should, if possible, reflect those costs.

The correlation matrix R is given in Table 1 and the rotated D matrix in Table 2. Note that $\Sigma\ d_{i1}^2 = 6.4$, while $d_{61}^2 = 6.15$. Hence row 6 defines a good regression with a residual sum of squares in the vertical norm not much greater than the best obtainable. The main predictors in this row are $X_4$, $X_5$, $X_6$, and $X_8$. Of these, $X_6$ is expensive. We now note that $X_6$ loads heavily on row 10, which also has a small $d_{10,1}$ value. Define transformation T by setting $t_{66} = 1/(1 + r^2)^{1/2}$, $t_{6,10} = -r/(1 + r^2)^{1/2}$, $t_{10,6} = r/(1 + r^2)^{1/2}$, and $t_{10,10} = 1/(1 + r^2)^{1/2}$, where $r + d_{6,6}/d_{6,10} = 0.04815$. This yields a new matrix $\hat{D} = TD$, given in Table 3. The residual sum of squares has virtually not increased, but variable 6 has been eliminated. A good regression subset is

**Table 1** Correlation Matrix R

| | | | | | | | | | |
|---|---|---|---|---|---|---|---|---|---|
| 1 | | | | | | | | | |
| 0.162 | 1 | | | | | | | | |
| 0.881 | 0.126 | 1 | | | | | | | |
| -0.883 | -0.123 | -0.998 | 1 | | | | | | |
| 0.213 | 0.137 | 0.234 | -0.233 | 1 | | | | | |
| 0.751 | 0.047 | 0.854 | 0.856 | 0.304 | 1 | | | | |
| 0.414 | 0.272 | 0.610 | -0.609 | 0.254 | 0.460 | 1 | | | |
| -0.392 | -0.300 | -0.610 | 0.612 | -0.176 | -0.504 | 0.968 | 1 | | |
| 0.182 | 0.121 | 0.144 | -0.138 | 0.546 | -0.157 | 0.343 | -0.216 | 1 | |
| 0.259 | 0.206 | 0.487 | -0.491 | -0.163 | 0.568 | 0.702 | -0.809 | -0.399 | 1 |

*Source:* Tables 1, 2, and 3 are courtesy of the Royal Statistical Society, England.

Table 2  Rotated D Matrix

| 0.00 | 0.00 | − 0.01 | − 0.66 | 0.03 | 0.48 | 0.00 | 0.10 | 0.00 | − 0.09 |
|---|---|---|---|---|---|---|---|---|---|
| 0.00 | 0.00 | 0.00 | − 0.72 | 0.72 | − 0.74 | 0.00 | 0.14 | − 0.16 | − 0.2 |
| 0.00 | 0.00 | 0.00 | 0.31 | 0.00 | − 0.29 | 0.00 | − 0.59 | − 0.06 | 0.69 |
| − 0.35 | 1.10 | − 1.08 | − 3.21 | − 1.02 | − 1.67 | 0.33 | 3.22 | 1.55 | 2.58 |
| 0.00 | 0.00 | 0.00 | − 0.39 | 0.29 | − 0.21 | 0.00 | 0.78 | − 0.78 | 0.51 |
| 2.48 | − 0.16 | 0.98 | 7.23 | 2.13 | 3.74 | − 0.16 | − 1.48 | − 0.69 | − 0.67 |
| − 0.09 | 0.08 | − 0.78 | − 1.37 | − 0.41 | − 0.63 | 4.81 | 5.81 | 0.22 | 1.41 |
| 0.11 | − 0.06 | 22.37 | 18.41 | − 2.11 | − 4.16 | − 0.16 | − 2.43 | − 1.59 | − 2.34 |
| − 0.08 | 0.10 | − 1.77 | − 4.00 | − 1.19 | − 2.18 | 0.40 | 59.01 | 37.50 | 62.39 |
| 0.32 | − 0.06 | 1.31 | 77.80 | 41.30 | 77.66 | − 0.10 | − 2.30 | − 1.18 | − 2.10 |

Table 3  Rotated D̂ Matrix

| 0.00 | 0.00 | − 0.01 | − 0.66 | 0.03 | 0.48 | 0.00 | 0.10 | 0.00 | − 0.09 |
|---|---|---|---|---|---|---|---|---|---|
| 0.00 | 0.00 | 0.00 | − 0.72 | 0.72 | − 0.74 | 0.00 | 0.14 | − 0.16 | 0.12 |
| 0.00 | 0.00 | 0.00 | 0.31 | 0.00 | − 0.29 | 0.00 | − 0.59 | − 0.06 | 0.69 |
| − 0.35 | 1.10 | − 1.08 | − 3.21 | − 1.02 | − 1.67 | 0.33 | 3.22 | 1.55 | 2.58 |
| 0.00 | 0.00 | 0.00 | − 0.39 | 0.29 | − 0.21 | 0.00 | 0.78 | − 0.78 | 0.51 |
| 2.46 | − 0.15 | 0.91 | 3.48 | 0.14 | 0.00 | − 0.16 | − 1.37 | − 0.63 | − 0.57 |
| − 0.09 | 0.08 | − 0.78 | − 1.37 | − 0.41 | − 0.63 | 4.81 | 5.81 | 0.22 | 1.41 |
| 0.11 | − 0.06 | 22.37 | 18.41 | − 2.11 | − 4.16 | − 0.16 | − 2.43 | − 1.59 | − 2.34 |
| − 0.08 | 0.10 | − 1.77 | − 4.00 | − 1.19 | − 2.18 | 0.40 | 59.01 | 37.50 | 62.39 |
| 0.44 | − 0.07 | 1.78 | 78.06 | 41.36 | 77.59 | 0.11 | − 2.37 | − 1.21 | − 2.13 |

therefore $X_4$, $X_5$, and $X_8$. As might be expected, in fitting the regression to these variables the coefficient of $X_5$ is not large. Thus a subset regression on $X_4$ and $X_8$ is tried and found to have a residual sum of squares of 0.187. This compares favorably with the residual sum of squares of 0.156 obtained when all variables are used. If there are two or more rows with large $d_{i1}$ values, this implies that perhaps the high loading variables of both (or their substitutes) may be needed in the final subset.

### 1.2.3. Logistic Regression

When the regression techniques above are discussed, it is usually presupposed that the dependent variable is continuous. The difficulties that arise when the dependent variable is binary are twofold. First, it is impossible for the error terms to be normally distributed. Second, the predicted values will not be binary, and one may obtain values that are less than zero or greater than 1. These values are difficult to interpret.

A different type of regression model overcomes some of these difficulties. This is the logistic model, in which it is assumed that one is observing a Bernoulli random variable Y taking value 0 with probability $1 - P$ and value 1 with probability P. This distribution for Y may be written compactly as $\text{Prob}(Y = y) = P^y(1 - P)^{1-y}$, where $y = 0$ or 1 as required.

The value P is unknown and is to be estimated. It is assumed that P is related to the independent variables $X_1, \ldots, X_m$ by the following relationship:

$$\log \frac{P}{1 - P} = \sum_{i=1}^{m} \beta_i X_i$$

where $\beta = (\beta_1, \beta_2, \ldots, \beta_m)^T$ is an unknown vector of coefficients. This implies that

$$P = \frac{\exp(\Sigma_{i=1}^{m} \beta_j X_j)}{1 + \exp(\Sigma_{i=1}^{m} \beta_j X_j)}$$

By estimating $\beta$ with $\hat{\beta}$, one obtains an estimate $\hat{P}$ for P, which is interpreted as the estimated probability that an individual with characteristics $X_1, \ldots, X_m$ will show $Y = 1$. In a pharmaceutical setting this corresponds to the estimated probability that a compound with physicochemical attributes $X_1, \ldots, X_m$ will show activity in a biological screen.

Let us now assume that n independent observations have been taken on the dependent variable associated with the n vectors $(X_{i1}, X_{i2}, \ldots, X_{im})$, $(i = 1, 2, \ldots, n)$. Denote the observed response for the ith vector by $y_i$. Then by the assumption of independence we have

$$\text{Prob}(Y_1 = y_1, \ldots, Y_n = y_n) = \prod_{i=1}^{n} \text{Prob}(Y_i = y_i)$$

$$= \prod_{i=1}^{n} P_i^{y_i}(1 - P_i)^{1-y_i}$$

Substituting

$$P_i = \frac{\exp(\Sigma_j \beta_j X_{ij})}{1 + \exp(\Sigma_j \beta_j X_{ij})}$$

we obtain the likelihood function

$$L(\beta) = \prod_{i=1}^{n} \left[ \frac{\exp(\Sigma_j \beta_j X_{ij})}{1 + \exp(\Sigma_j \beta_j X_{ij})} \right]^{y_i} \left[ \frac{1}{1 + \exp(\Sigma_j \beta_j X_{ij})} \right]^{1-y_i}$$

$$= \prod_{i=1}^{n} \left[ \frac{\exp(y_i \Sigma_j \beta_j X_{ij})}{1 + \exp(\Sigma_j \beta_j X_{ij})} \right]$$

We now find $\beta$ that maximizes $L(\beta)$. This requires an iterative procedure. One possibility is the stepwise logistic regression program which has recently been incorporated in the BMD series. Like the regression procedures discussed in the previous sections, the independent variables $X_1, \ldots, X_m$ may be either continuous, binary, or a mixture of both.

For pharmaceutical data with an active-inactive response variable, logistic regression may be a valuable QSAR technique. Additional discussion on this topic may be found in Haberman (1974).

## 1.3. CLASSIFICATION

A problem closely related to that of discovering active compounds is the problem of classification. It may be described as follows. Suppose that there are two populations, A and B, and two groups of individuals, one known to belong to population A and the other to population B. Suppose now that an unknown individual appears and is to be classified as belonging to either A or B on the basis of some observed characteristics. How should we use the information about the observed characteristics of the known individuals to help us correctly classify this unknown individual? When we let A denote the population of active compounds, B the population of inactive compounds, and let the observed characteristics of a compound be its physicochemical attributes, the relevance of this classification problem to pharmaceutical research becomes immediately apparent.

In this section we describe several classification procedures. Included among them are some recent relatively unknown methods that show promise of being useful.

We begin to build our conceptual framework for classification problems by considering one of the simplest classification schemes. This is the nonparametric k-nearest-neighbors procedure. Then we will describe the linear learning machine algorithm, which finds a separating hyperplane between two linearly separable populations. This nonparametric pattern recognition procedure has received considerable attention recently in the QSAR literature.

A different and parametric approach to the classification problem may be developed within the decision-theoretic framework of Wald. A general description of this next framework is given, and as an illustration Fisher's discriminant function is derived. Then a more recent procedure by Day and Kerridge is described. This demonstrates that the parametric approach may lead to classification procedures which possess some of the features sought after in the nonparametric schemes without some of the accompanying pitfalls. In particular, their logistic discriminant functions will define a separating hyperplane if one exists, and will be reasonably behaved if one does not. Finally, we turn our attention to kernel methods, the least well known of these classification procedures, yet among the most promising. While they arise out of the parametric tradition, they are essentially nonparametric in their assumptions and appear to have potentially wide applications.

### 1.3.1. k Nearest Neighbors

Undoubtedly, the simplest of all classification schemes is the k-nearest-neighbors method. It was introduced in a technical report by Fix and Hodges (1951), but because of its simplicity did not receive much initial attention. Now, however, it has become a standard technique described in most pattern recognition texts (see, e.g., Andrews, 1972).

The context is the one described in the introduction. We assume that there are two populations, A and B, and we are to classify each new individual on the basis of a vector of observed characteristics $y = (y_1, \ldots, y_m)$ as belonging to population A or population B. To aid us in making our decision, we have a training set T, consisting of individuals previously classified correctly, together with their observed feature vectors. In addition, we have a metric $d(x,y)$ on the feature space which measures the similarity between any two feature vectors. Exactly how this metric is defined will vary according to what we may feel to be important in measuring similarity. However, for real-valued measurements the obvious metric is Euclidean distance,

$$d(x,y) = \left[ \sum_{i=1}^{m} (x_i - y_i)^2 \right]^{1/2}$$

while for binary present/absent data it is matched pairs, that is,

$$d(x,y) = \text{number of matched pairs}$$

$$= \sum_{i=1}^{m} \delta(x_i, y_i)$$

where

$$\delta(x_i, y_i) = \begin{cases} 1 & \text{if } x_i = y_i \\ 0 & \text{if } x_i \neq y_i \end{cases}$$

With this context, a few lines will now serve to describe the k-nearest-neighbors method. Given a new individual with observed feature vector y, find the k nearest feature vectors already classified, then assign the individual to the class representing the majority of those k nearest vectors. Usually, k is chosen to be an odd number to prevent ties.

Interest in the k-nearest-neighbors classification method may be attributed to two elements. First, it makes no assumptions about the underlying populations. Second, it does reasonably well in practice. Cover and Hart (1967) obtained some theoretical results supporting this performance record. They showed that if R is the probability of misclassification using the optimal classification rule when the underlying population distributions are known, and representable as a mixture of discrete and continuous distributions, and $R_n$ is the error rate using the one-nearest-neighbor rule with a training set of n individuals, then as n becomes large, one obtains in the limit

$$\hat{R} \leqslant R_\infty \leqslant 2\hat{R}(1 - \hat{R})$$

Thus, if $\hat{R} = 0.050$, then

$$0.050 \leqslant R_\infty \leqslant 0.095$$

How large n must be to achieve error rates nearly within these bounds cannot of course be determined a priori. In practice, several hundred individuals must already be correctly classified to obtain moderate accuracy. To achieve a high success rate, many more are needed. It is at this point that one of the weaknesses of the k-nearest-neighbors method becomes apparent. The rule requires that the distance between any new individual and each of the classified ones be computed and compared to find the k nearest. If n is large, this implementation phase of the rule is expensive.

A procedure rather similar to the k-nearest-neighbors rule is the kernel method, to be discussed later. Because of its construction, which makes more effective use of all the information available in the training set, one would anticipate that the kernel procedures would be superior. However, there is no empirical evidence yet to support this claim.

## 1.3.2. Linear Learning Machine

A more sophisticated nonparametric method that has received considerable attention in pattern recognition texts is the linear learning machine algorithm. This technique has also received widespread attention in the QSAR literature (see, e.g., Isenhour and Jurs, 1973, and Stuper et al., 1979).

The problem addressed by the technique is the following. Let $T_A$ and $T_B$ be the feature vectors associated with the training sets of individuals previously correctly classified as belonging to populations A and B, respectively. We shall say that $T_A$ and $T_B$ are linearly separable in the feature space V if there exists a linear function $g(y) = \beta_0 + \Sigma_{i=1}^m \beta_i y_i$ of the coordinates of any feature vector $y = (y_1, \ldots, y_m)$ in V such that $g(y) > 0$ whenever $y \in T_A$ and $g(y) < 0$ whenever $y \in T_B$. The existence of such a function is equivalent to the existence of a hyperplane separating sets $T_A$ and $T_B$. It is then natural to classify a new individual as belonging to either population A or B, depending on which side of the dividing hyperplane its feature vector lies. In terms of the function g, this is equivalent to classifying an individual with vector y as belonging to A if and only if $g(y) > 0$. This defines a simple-to-administer rule with appealing geometrical properties (see Figure 7).

The linear learning machine algorithm has the property that it finds such a function if one exists. To describe the procedure, it is convenient to reparameterize the space V as follows. For each $y \in V$ define $\tilde{y}$ by $\tilde{y} = (1, y_1, \ldots, y_m)^T \in \tilde{V}$, where $\tilde{V}$ is V augmented with one additional dimension. Thus any linear function $g(y)$ may be conveniently written as the vector product $\beta^T \tilde{y}$, where $\beta^T = (\beta_0, \ldots, \beta_m)$ is the set of coefficients defining g. Now define $\tilde{T}_A$ to be the set in $\tilde{V}$ corresponding to $T_A$ in V, and define

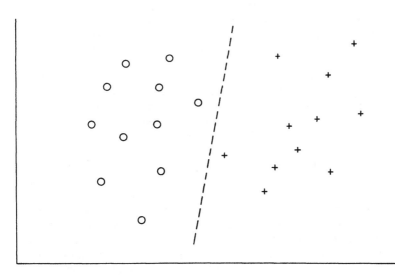

Figure 7  Separating hyperplane.

$\tilde{T}_B$ to be the negative of the set corresponding to $T_B$, that is, $\tilde{T}_B = \{\tilde{y} \in \tilde{V} : \tilde{y}T = -(1, y^T), y \in T_B\}$. Thus finding a linear function g with the property $g(y) > 0$ whenever $y \in T_A$ and with $g(y) < 0$ whenever $y \in T_B$ is equivalent to finding a vector $\beta$ such that $\beta^T \tilde{y} > 0$ for all $\tilde{y} \in \tilde{T}_A \cup \tilde{T}_B$.

To begin the algorithm, arrange the vectors in $\tilde{T}_A \cup \tilde{T}_B$ in some order, yielding a sequence $\tilde{z}_1, \ldots, \tilde{z}_n$, where n is the total number of individuals in the training sets. By repeating this sequence an infinite number of times we obtain the infinite sequence $\{\tilde{z}_1\}_{i=1}^{\infty}$, where $\tilde{z}_{i+n} = \tilde{z}_i$ for all i. Then set the vector of coefficients $\beta^{(0)}$ to some arbitrary set of values and begin to generate a sequence $\{\beta^{(i)}\}_{i=0}^{\infty}$ by the following recursive rule:

$$
\beta^{(i+1)} = \begin{cases} \beta^{(i)} & \text{if } \beta^{(i)T} \tilde{z}_{i+1} > 0 \\[2mm] \beta^{(i)} + c\tilde{z}_{i+1} & \text{if } \beta^{(i)T} \tilde{z}_{i+1} < 0 \end{cases}
$$

where $c < 0$ is typically chosen to be $2\beta^{(i)T}\tilde{z}_{i+1}/\tilde{z}_{i+1}^T\tilde{z}_{i+1}$ for its good convergence properties. Stop at stage k of this procedure if the last n $\beta$'s did not change [i.e., $\beta^{(k)} = \beta^{(k-1)} = \cdots = \beta^{(k-n+1)}$]. Then $\beta^{(k)}$ is the vector we sought.

The learning machine may be used on any feature space that may be embedded in a vector space endowed with an inner product. In particular, it may be used with binary features when such vectors of zeros and ones are considered ordinary Euclidean vectors.

As portrayed in Figure 7, the linear learning machine algorithm clearly appears to be an attractive classification procedure. However, several undesirable features of this technique should be noted. First, if a separating function exists, there will generally be infinitely many such functions, and the algorithm does not discriminate between them, simply picking the first one it finds. The geometrical implications of this are illustrated in Figure 8. On intuitive grounds one may feel that the dashed hyperplane would serve as a better separating boundary than the solid hyperplane. In particular, one may prefer to assign the unknown individual marked □ as belonging to the ○ population rather than the + population. However, if the solid hyperplane is found first, that will be the separating hyperplane specified by the algorithm.

A second, even more serious drawback is the algorithm's behavior when a separating hyperplane does not exist. In that case the algorithm as given never stops! This case may in fact

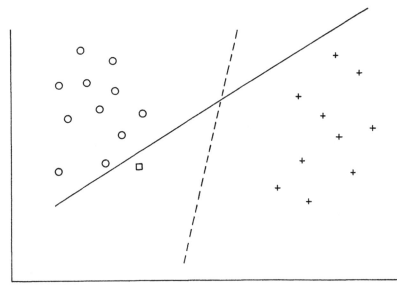

**Figure 8**  Two separating hyperplanes.

be typical in pharmaceutical research. In the $\pi$-$\sigma$ plane, Hansch's parabolic regression equation leads one to expect a local cluster of actives surrounded by inactives. This difficulty may be overcome in part by taking advantage of an observation by Minsky and Papert (1969) that the sequence $\{\beta^{(i)}\}_{i=0}^{\infty}$ will begin to cycle through a sequence of vector values after some stage if no separating hyperplane exists. Consequently, the basic algorithm may be modified to detect cycling and to stop whenever it occurs, However, the required number of steps before termination in such circumstances will at least be twice the number of observations. If there are many observations, the effort expended could be large.

The drawbacks listed above are serious. When it is simultaneously realized that there exist parametric procedures which also find separating hyperplanes when they exist without the pitfalls, considerable doubt is cast upon the usefulness of the learning algorithm for the type of classification problems considered here.

It would, however, not be too difficult to modify the algorithm so as (ii) to achieve a best separating hyperplane in some sense, thereby avoiding the situation illustrated in Fig. 8, and (ii) to find nearby separating hyperplanes, when no strictly separating hyperplane exists, or can be found in an acceptable time. An algorithm with these additional features would be worthy of serious consideration.

### 1.3.3. Decision-Theoretic Formulation of the Classification Problem

A more sophisticated conceptual framework for the classification problem may be gained through the decision-theoretic approach. Ideally, we would like to find a procedure that will permit us to classify an unknown individual correctly every time. Typically, we shall not be able to find such a procedure, and consequently, for any procedure we might consider, there are apt to be two kinds of misclassification errors. First, an active individual may be classified as inactive, and second, an inactive may be classified as active. The seriousness of these two types of error may not be equal, and therefore we assign to them two different cost, $c_1$ and $c_2$. Given these costs, we may define our goal as one of finding the procedure that will minimize our average cost of miclassification.

Let us now consider any arbitrary classification rule d which assigns population membership on the basis of observed features. With two populations this is equivalent to partitioning the feature space into two mutually exclusive regions R and $R^c$ such that if $y \in R$ is observed, the individual is classified as belonging to A, while if $y \notin R$, he is classified as belonging to B. Hence the task of finding an optimal rule d is equivalent to finding an optimal partition R and $R^c$.

To get a theoretical handle on the problem, we shall now assume that there exist probability measures $P(S|A)$ and $P(S|B)$ which describe the probability of observing an individual with a feature vector belonging to subset S of the feature space V, given that this individual comes from population A or B, respectively. In addition, we assume that there exist prior probabilities $P(A)$ and $P(B)$ [with $P(A) + P(B) = 1$] for the events that the next individual presented for classification will in fact belong to A or B.

It now follows that the expected cost of using the partition (R and $R^c$) as a decision rule is

$$c_1 P(R^c|A)P(A) + c_2 P(R|B)P(B) \qquad (*)$$

where, for example, the first term arises from the possibility of wrongly classifying an individual as belonging to population B when he or she in fact belongs to population A. Letting $f_A(y)$ and $f_B(y)$ be the density functions corresponding to $P(S|A)$ and $P(S|B)$, we may write $(*)$ as follows:

$$c_1 P(A) \int_{R^c} f_A(y)\, dy + c_2 P(B) \int_R f_B(y)\, dy$$

Noting that

$$\int_{R^c} f_A(y)\ dy = 1 - \int_R f_A(y)\ dy$$

we have

$$c_1 P(A) + \int_R [c_2 P(B) f_B(y) - c_1 P(A) f_A(y)]\ dy$$

This quantity is clearly minimized when we integrate over the region for which the integrand is negative, that is, the region

$$R^* = \{y : c_2 P(B) f_B(y) \leqslant c_1 P(A) f_A(y)\}$$

$$= \left\{ y : \frac{f_A(y)}{f_B(y)} \geqslant \frac{c_2 P(B)}{c_1 P(A)} \right\}$$

Consequently, the boundary $\partial R^*$ separating the two regions consists of those values of y for which

$$c_1 P(A) f_A(y) = c_2 P(B) f_B(y)$$

Using the relationship

$$f_A(y) = \frac{P(A|y) f(y)}{P(A)}$$

we may also describe the optimal region as

$$R^* = \left\{ y : P(A|y) \geqslant \frac{c_2}{c_1} P(B|y) \right\}$$

Thus, when $c_1 = c_2$, we see that an individual with feature vector y should be classified as belonging to A if and only if the posterior probability of A given y is greater than the posterior probability of B given y. This is as we might have expected.

At this point a word is in order about the meaning of probabilities such as $P(A)$ and $P(S|A)$. These refer to the population of compounds which the scientist concerned considers worthy of testing. If he or she were able to specify this population in advance of any experimentation, and also undertook to select compounds for testing in a controlled random manner, these probabilities would have precise meanings. In practice, neither of these two conditions is likely to be fulfilled, and it is a matter of judgment and experience to decide whether our model remains a useful one. Fortunately, its usefulness does not depend on its precise validity. This is simply a question of whether it tends to classify compounds correctly as active or inactive, and can be answered empirically. Even if the assumed parametric form of the model is inaccurate, it may still do a good job of classification. Typically, the parameters of the model, on which some of the relevant probabilities depend, must be estimated from the data as testing proceeds.

### 1.3.4. Fisher's Discriminant Function

The best known and most widely used parametric classification procedure is Fisher's discriminant function. A description of it may be found in many texts (see, e.g., Maxwell, 1977, and Timm, 1975).

Suppose initially that $c_1 = c_2$, $P(A) = P(B)$, and $f_A(y)$ and $f_B(y)$ are both multivariate normal density functions with common covariance matrix $V$ and respective means $\mu_A$ and $\mu_B$. For example, $f_A(y)$ has the form

$$f_A(y) = \frac{\exp[-(1/2)(y - \mu_A)^T V^{-1}(y - \mu_A)]}{(2\pi)^{m/2}|V|^{1/2}}$$

Upon taking logs, we see that a point $y$ on the optimal boundary

$$\partial R^* = \left\{ y : \frac{f_A(y)}{f_B(y)} = 1 \right\}$$

satisfies

$$-\frac{1}{2}(y - \mu_A)^T V^{-1}(y - \mu_A) = -\frac{1}{2}(y - \mu_B)^T V^{-1}(y - \mu_B)$$

Denoting the elements of $V^{-1}$ by $\alpha_{ij}$ and canceling and collecting terms, this may be written as

$$\sum_i \sum_j \alpha_{ij} (\mu_{Ai} - \mu_{Bi}) y_j = \sum_i \sum_j \alpha_{ij} (\mu_{Ai} \mu_{Bi} - \mu_{Aj} \mu_{Bj})$$

where $\mu_{Ai}$ is the ith component of the mean vector $\mu_A$. Our optimal classification rule is therefore: Compute

$$D(y) = \sum_i \sum_j \left[ \alpha_{ij} (\mu_{Ai} - \mu_{Bi}) - \frac{1}{2} (\mu_{Ai} \mu_{Bi} - \mu_{Aj} \mu_{Bj}) \right]$$

and classify the individual with feature vector y as belonging to population A if $D(y) \geq 0$ and as belonging to population B if $D(y) < 0$. We may now note two things. First, $D(y)$ is a linear function of the components $y_i$ $(i = 1,2,\ldots,m)$. Second, its computation depends on knowing $\mu_A$, $\mu_B$, and V. To make this dependence explicit we denote $D(y)$ by $D(y|\mu_A,\mu_B,V)$. In general, we do not know $\mu_A$, $\mu_B$, and V, and therefore it is not possible to compute $D(y|\mu_A,\mu_B,V)$.

Let us now assume that there is a training set $R = T_A \cup T_B$, consisting of $n_A$ individuals known to come from population A and $n_B$ known to come from population B. We can use this information to estimate the unknowns $\mu_A$, $\mu_B$, and V, and construct a classification function based on these estimates. From multivariate normal theory the maximum likelihood estimators for $\mu_A$, $\mu_B$, and V are, respectively, $\bar{y}_A$, $\bar{y}_B$, and S, where $\bar{y}_A$ is the sample mean of $T_A$, $\bar{y}_B$ is the sample mean of $T_B$, and S is the pooled sample covariance matrix. This yields Fisher's discriminant function $D(y|\bar{y}_A,\bar{y}_B,S)$.

This same classification function may be obtained by a nonparametric argument. Suppose that we seek a linear function

$$\ell(y) = \beta_0 + \sum_{j=1}^{m} \beta_j y_j$$

of the components of y such that the ratio of the between-class sample variance to the within-class sample variances of the training sets is maximized along $\ell$. This is a standard technique used to separate two distinguishable groups as much as possible along a hyperplane in the feature space. The quantity to be maximized is therefore

$$\frac{[(1/n_A)\Sigma_i \mathcal{L}(y_A^i) - (1/n_B)\Sigma_i \mathcal{L}(y_B^i)]^2}{[1/(n_A + n_B)]\{S_i[\mathcal{L}(y_A^i) - (1/n_A)\Sigma_j \mathcal{L}(y_A^j)]^2 + \Sigma_i[\mathcal{L}(y_B^i) - (1/n_B)\Sigma_j \mathcal{L}(y_B^j)]^2\}}$$

$$= \frac{\Sigma_i \Sigma_j \beta_i \beta_j (\bar{y}_{Ai} - \bar{y}_{Bi})(\bar{y}_{Aj} - \bar{y}_{Bj})}{\Sigma_i \Sigma_j \beta_i \beta_j S_{ij}}$$

where

$$S_{ij} = \frac{\Sigma_i \Sigma_j (y_{Aj}^i - \bar{y}_{Aj})^2 + \Sigma_i \Sigma_j (y_{Bj}^i - \bar{y}_{Bj})^2}{n_A + n_B}$$

and $y_A^i$ is the ith vector belonging to training set $T_A$. Differentiating with respect to $\beta_k$ and setting the result equal to zero, we obtain the m equations

$$\bar{y}_{Ak} - \bar{y}_{Bk} = \frac{\Sigma_i[\beta_i(\bar{y}_{Ai} - \bar{y}_{Bi})\Sigma_j S_{kj}\beta_k]}{\Sigma_i \beta_i \beta_j S_{ij}} \qquad (k = 1, 2, \ldots, m)$$

These may be solved in $\beta_k$ to obtain

$$\beta_k = C \sum_{j=1}^m \gamma_{jk}(\bar{y}_{Aj} - \bar{y}_{Bj})$$

where C is a constant and $\gamma_{ij}$ is the (i,j)th element of the inverse $\Gamma$ to S. Since multiplying all the $\beta_k$ by a constant does not alter the direction of the line , the coefficient C is usually set equal to 1. Hence an optimal discriminant line is

$$*(y) = \sum \sum \gamma_{ji}(\bar{y}_{Aj} - \bar{y}_{Bj})y_i + \text{constant}$$

This is equivalent to $D(y|\bar{y}_A, \bar{y}_B, S)$.

In deriving Fisher's discriminant function it has been assumed that the covariances of populations A and B are equal. This is an assumption that may be tested. When it is rejected, the general form of the boundary $\partial R*$ becomes

$$\frac{1}{2} \log |V_A| - \frac{1}{2} \sum_i \sum_j \alpha_{Aij} (y_i - \mu_{Ai})(y_j - \mu_{Aj})$$

$$= \frac{1}{2} \log |V_B| - \frac{1}{2} \sum_i \sum_j \alpha_{Bij} (y_i - \mu_{Bi})(y_j - \mu_{Bj})$$

which is quadratic, containing cross-product terms $y_i$ and $y_j$ ($i, j = 1, 2, \ldots, m$). From this a quadratic classification function may be obtained by substituting the appropriate maximum likelihood estimators for $\mu_A$, $\mu_B$, $V_A$, and $V_B$.

Without difficulty one may also include different costs of misclassification, and different prior probabilities of observing an individual from population A or from B. This does not change the form of the classification boundary, which is given by the equation

$$\frac{1}{2} \log |V_A| - \frac{1}{2} \sum_i \sum_j \alpha_{Aij} (y_i - \mu_{Ai})(y_j - \mu_{Aj})$$

$$= \frac{1}{2} \log |V_B| - \frac{1}{2} \sum_i \sum_j \alpha_{Bij} (y_i - \mu_{Bi})(y_j - \mu_{Bj}) + \log \frac{c_2 P(B)}{c_1 P(A)}$$

### 1.3.5. Logistic Classification

Day and Kerridge (1967) have generalized the Fisher discriminant through the following interesting observation. Assume that the multi-variate densities for populations A and B are of the following form:

$$f(y \,|\, A) = a_A \exp \left[ -\frac{1}{2} (y - \mu_A)^T V^{-1} (y - \mu_A) \right] \phi(y)$$

$$f(y \,|\, B) = a_B \exp \left[ -\frac{1}{2} (y - \mu_B)^T V^{-1} (y - \mu_B) \right] \phi(y)$$

where $a_A$ and $a_B$ are chosen such that these densities integrate to 1, and where $\phi(y)$ is an arbitrary positive function of $y$. Then the optimal boundary is all $y$ such that

$$-\frac{1}{2} (y - \mu_A)^T V^{-1} (y - \mu_A) = -\frac{1}{2} (y - \mu B)^T V^{-1} (y - \mu_B) + \log \frac{a_B c_2 P(B)}{a_A c_1 P(A)}$$

That is, $\partial R^*$ does not depend on $\phi(y)$! Since $\phi(y)$ is arbitrary, $f(y \,|\, A)$ can take many different forms, with $f(y \,|\, B)$ differing from

$f(y|A)$ within the range of possibilities permitted by varying $\mu_A$, $\mu_B$, and V. The Fisher discriminant function may be applied in all these cases, as long as the underlying distributions are completely known.

As Day and Kerridge point out, the above does not imply that we will obtain the function $D(y|\bar{y}_A, \bar{y}_B, S)$ when $\mu_A$, $\mu_B$, and V are unknown. The argument used in obtaining the function above previously was that $\bar{y}_A$, $\bar{y}_B$, and S were the maximum likelihood estimators for $\mu_A$, $\mu_B$, and V when the density functions were multivariate normal. If $\phi(y) \neq 1$, this is no longer the case and thus the same results usually will not be obtained.

For this more general case, where $\mu_A$, $\mu_B$, and V are unknown and $\phi(y) \neq 1$, but known, Day and Kerridge look directly at the likelihood function. It is

$$L = \prod_{y \in T_A} f(y|A)P(A) \times \prod_{y \in T_B} f(y|B)P(B)$$

$$= \prod_{y \in T_A} a_A P(A) \phi(Y) \exp\left[-\frac{1}{2}(y - \mu_A)^T V^{-1}(y - \mu_A)\right]$$

$$\times \prod_{y \in T_B} a_B P(B) \phi(y) \exp\left[-\frac{1}{2}(y - \mu_B)^T V^{-1}(y - \mu_B)\right]$$

To obtain the maximum likelihood estimators for $\mu_A$, $\mu_B$, and V is now simply a matter of maximizing L with respect to these parameters. How difficult this is depends on the form of $\phi(y)$. If $\hat{\mu}_A$, $\hat{\mu}_B$, and $\hat{V}$ are the resulting estimates, we obtain the discriminant function $D(y|\hat{\mu}_A, \hat{\mu}_B, \hat{V})$.

However, Day and Kerridge go far beyond this simple extension and consider the much more theoretically difficult problem of how to develop a classification function within this framework when $\phi(y)$ is unknown. This case is of considerable interest, for in permitting $\phi$ to be unknown they are essentially permitting the density functions to be unknown in form, and thus are effectively considering a nonparametric classification problem. To get a handle on this problem, Day and Kerridge rewrite the likelihood function as

$$L = \prod_{y \in T_A \cup T_B} f(y) \prod_{y \in T_A} P(A|y) \prod_{y \in T_B} P(B|y)$$

where $f(y) = f(y|A)P(A) + f(y|B)P(B)$, the unconditional density of y. Using the relation

$$P(A|y) = \frac{f(y|A)P(A)}{f(y|A)P(A) + f(y|B)P(B)}$$

we note after some simplification that

$$P(A|y) = \frac{\exp(y^T c + d)}{1 + \exp(y^T c + d)}$$

where $c = V^{-1}(\mu_A - \mu_B)$, a vector, and

$$d = -\frac{1}{2}(\mu_A - \mu_B)^T V^{-1}(\mu_A + \mu_B) + \log \frac{P(A)a_A}{P(B)a_B}$$

a scalar. That is, the posterior probability of A is a logistic function with argument $y^T c + d$. Substituting this expression into the likelihood, together with the appropriate expression for $f(y)$, the likelihood becomes

$$L = \prod_{y \in T_A \cup T_B} \phi(y) \left\{ a_A P(A) \exp\left[ -\frac{1}{2}(y - \mu_A)^T V^{-1}(y - \mu_A) \right] \right.$$

$$\left. + a_B P(B) \exp\left[ -\frac{1}{2}(y - \mu_B)^T V^{-1}(y - \mu_B) \right] \right\}$$

$$\times \prod_{y \in T_A} \frac{\exp(y^T c + d)}{1 + \exp(y^T c + d)} \prod_{y \in T_B} \frac{1}{1 + \exp(y^T c + d)}$$

This expression must be maximized by suitable choice of $\phi(y)$, $P(A)$, $P(B)$, $M_A$, $M_B$, and V. We may regard the maximization as being first with respect to c and d, with a subsequent maximization with respect to $\phi(y)$, $P(A)$ and $P(B)$ for fixed c and d, affecting only the factors included in the first product over $T_A \cup T_B$. Day and Kerridge argue that the maximum value of this product depends only weakly on the values of c and d. Thus we may in practice disregard this factor when trying to maximize the likelihood. When we do disregard this factor we obtain the reduced likelihood

$$L = \prod_{y \in T_A} \frac{\exp(y^T c + d)}{1 + \exp(y^T c + d)} \prod_{y \in T_B} \frac{1}{1 + \exp(y^T c + d)}$$

It is to be maximized in c and d. Equivalently, we may maximize the logarithm of the reduced likelihood function

$$\log L = n_A(\bar{y}_A c + d) - \sum_{y \in T_A \cup T_B} \log[1 + \exp(y^T c + d)]$$

in these same parameters. Typically, this will be done by an iterative hill-climbing algorithm. Once the estimates $\hat{c}$ and $\hat{d}$ are obtained, y is classified as belonging to A if $\hat{P}(A|y) \geq \hat{P}(B|y)(c_2/c_1)$, where $c_1$ and $c_2$ are the costs of misclassification.

An interesting property of the foregoing solution to the classification problem is that like the linear learning machine, it will define a separating hyperplane between groups $T_A$ and $T_B$ if they are linearly separable. To see this, let $g(y) = d + y^Tc$ be an arbitrary linear classification function. Without loss of generality we may assume that $g(y)$ classifies y as belonging to population A if and only if $g(y) > 0$. The reduced likelihood function $L(g)$ may be written

$$L(g) = \prod_{y \in T_A} \frac{e^{g(y)}}{1 + e^{g(y)}} \prod_{y \in T_B} \frac{e^{-g(y)}}{1 + e^{-g(y)}}$$

since

$$\frac{e^{-g(y)}}{1 + e^{-g(y)}} = \frac{1}{1 + e^{g(y)}}$$

Each term in the product takes a value between 1/2 and 1 if the corresponding y is correctly classified, and a value between 0 and 1/2 if misclassified.

Now let g* be a linear discriminant which correctly classifies each y in $T_A$ and $T_B$. Thus $g^*(y) > 0$ for y in $T_A$ and $g^*(y) < 0$ for y in $T_B$, and consequently

$$L(\theta g^*) \to 1$$

as $\theta \to \infty$. Hence, for some $\theta^*$,

$$L(\theta^*g^*) > \frac{1}{2}$$

This implies that as L is maximized, one will obtain a function g' at some stage for which $L(g') > 1/2$, if a separating g* exists. Moreover, g' is itself separating, since any nonseparating g has a factor which is less than 1/2, and consequently $L(g) < 1/2$. Note, further, that the existence of a separating g* implies, by the argument above, that no maximizing $\hat{c}$ and $\hat{d}$ exist.

The logistic classification procedure may also be derived by a different line of argument. One could assume that the underlying population has a logistic density with unknown parameters c and d. This again yields the classification function $y^Tc + d$. This is of some interest, for it implies that classification on the basis of the model above is essentially the same as classification based on the log-linear model with no interaction terms (see, e.g., Bishop et al., 1975).

One of the strengths of the logistic classification function is that it may handle both categorical and continuous feature vectors. In addition, it may be generalized to include quadratic and higher-order terms. Anderson (1975) has investigated this possibility.

Apparently, this model has yet to receive attention in the pharmaceutical literature. In view of its promise, it certainly warrants some consideration.

## 1.3.6. Kernel Methods

Very promising classification procedures have recently been proposed by Habbema et al. (1974) and Aitchinson and Aitken (1976). These are kernel-based techniques, and they represent a further evolution of the parametric approach toward essentially nonparametric assumptions.

As we have seen, a natural classification procedure is easily obtained once one is able to estimate the density functions $f(y|A)$ and $f(y|B)$. The kernel method proposes that these [e.g., $f(y|A)$] be estimated by a function of the form

$$f(y|A,\lambda) = \frac{1}{n_A} \sum_{i=1}^{n_A} K(y|y^i,\lambda) \tag{*}$$

where the sum is taken over all vectors belonging to the training set $T_A$. The kernel $K(y|y^i, \lambda)$ is chosen to be a function of simple form which corresponds to the notion of placing a small sand hill of probability around the observed feature point $y^i$ in the feature space V. The spread of this sand hill is determined by the parameter $\lambda$. This implies we estimate the density at a new point y by the sum of such sand hills, the influence of any one decreasing with distance from this new point. Consequently, we obtain a density function with hills and valleys corresponding to varying degrees of concentration of the points found in the training set.

For the continuous case, Habbema et al. propose a normal kernel of the form

$$K(y|x, \lambda) = \frac{\exp[-(1/2)(y - x)^T(y - x)/\lambda]}{(2\pi\lambda)^{m/2}}$$

This is spherically symmetric, but clearly it may be modified by incorporating the $T_A$ sample correlation matrix R to obtain

$$K(y|x, \lambda, R) = \frac{\exp[(1/2)(y - x)^T R^{-1}(y - x)/\lambda]}{(2\pi\lambda)}$$

The parameter $\lambda$ has then to be judicously chosen: not too small, in which case the resulting density is too peaked and sample dependent; and not too large, resulting in a density function that is too uniform. A successful method suggested by Habbema is to use what is known as a jackknife likelihood technique, in which $\lambda_A$ is chosen for the training set $T_A$ to maximize

$$W(\lambda_A|T_A) = \prod_{i=1}^{n_A} f_i(y_i|\lambda_A)$$

where

$$f_i(y_i|\lambda_A) = \frac{1}{n_A - 1} \sum_{j \neq 1} K(y_i|y^j, \lambda_A)$$

In effect, $\lambda_A$ is chosen such that we are best able to predict each member of the training set using only the presence of the others. This is a maximization problem in only one variable, and therefore a relatively simple task to accomplish.

Once $\lambda_A$ is obtained for training set $T_A$, and similarly $\lambda_B$ is obtained for training set $T_B$, the implementation of the classification rule is straightforward. For any new vector y, compute $f(y|A,\lambda_A)$ by summing $K(y|y^i,\lambda)$ over all $y^i$ belonging to $T_A$. Similarly, compute $f(y|B,\lambda_B)$. Then classify y as belonging to population A if and only if

$$\frac{f(y|A,\lambda_A)}{f(y|A,\lambda_B)} \geqslant \frac{c_2 P(B)}{c_1 P(A)}$$

Aitchison and Aitken have extended the continuous kernel method to both the binary and the mixed binary-continuous cases. For the binary case they suggest a kernel of the form

$$K(y|x,\lambda) = \lambda^{k-d(x,y)}(1 - \lambda)^{d(x,y)}$$

where $1/2 \leqslant \lambda \leqslant 1$ and $d(x,y)$ is the number of mismatches in the vectors x and y. For a given data point x, the contribution of the kernel to the density function is therefore to place density $\lambda^k$ on x itself and $\lambda^{k-h}(1 - \lambda)^h$ on each of the $m!/h!(m - h)!$ points of $B^m$ (the binary feature space of dimension m), differing in exactly h components from x. It may then be noted that the extreme case $\lambda = 1/2$ corresponds to putting an equal mass on all y points and hence is independent of the data, while $\lambda = 1$ corresponds to putting unit mass on y = x, and on no other y for the given x. Thus a sensible $\lambda$ would fall somewhere between 1/2 and 1.

The jackknife likelihood method of Habbema may also be applied to this type of kernel to obtain a good choice for $\lambda$. With this done, the classification procedure is then identical to the continuous case discussed previously.

Aitchison and Aitken prove that the kernel method has good consistency properties in the binary case. In particular, if $\lambda_n$ is estimated by the jackknife likelihood method as a function of a training set of n data points, then $f(y|A,\lambda_n)$ convertes in probability to the true distribution $f(y|A)$ for each $y \in B^m$ as $n \to \infty$. This implies that the kernel method will yield a classification rule which is asymptotically optimal.

Going beyond this aspect of binary discrimination, Aitchison and Aitken have suggested an "atypicality" index which will indicate to what degree an unknown individual is atypical of a specific training set. Thus if an unknown is atypical of both training sets, it may be best not to assign him to any population, and assume that he is not classifiable on the basis of the current training sets. This atypicality measure is obtained as follows. A feature vector y will be termed more typical of training set $T_A$ than vector z if

$$f(y|A,\lambda) > f(z|A,\lambda)$$

For given z, compute the total probability assigned to all binary vectors more typical of $T_A$ than z by the given fitted density for population A. That is, set

$$I_z^A = \sum_y f(y|A,\lambda)$$

where the sum is taken over all y such that y is more typical than z of $T_A$. This is computationally feasible when the dimension m is moderate. (i.e., $m \leqslant 10$). Now, if $I_z^A$ and $I_z^B$ are near 1, z is atypical for both $T_A$ and $T_B$, and hence is not predictable on the basis of these training sets.

The kernel method may be generalized to handle the parametrically awkward problem of discrimination with mixed binary and continuous features. In particular, let the data for one training set be

$$D_A = \{(b_i,c_i); b_i \in B^m, c_i \in R^\ell\}$$

when $b_i$ is binary and $c_i$ is continuous. For (y,z), a typical vector in $B^m \times R^\ell$, the following product kernel is used:

$$f(y,z|A,\lambda,\delta) = \frac{1}{n}\sum_{i=1}^{n} K(y|A,\lambda)L(z|A,\delta)$$

where K is the binary kernel suggested by Aitchison and Aitken and L is the normal kernel suggested by Habbema. Then both $\lambda$ and $\delta$ are estimated by the jackknife likelihood method. Note that the fact that the kernel is the product of two pseudolikelihood functions does not imply that the binary and continuous components of the feature vector are independent.

The extension above demonstrates one of the strengths of the kernel method, namely, its ability to describe complex dependence in terms of extremely simple functions. It is essentially a non-parametric method which permits the data points themselves to fashion any dependences or interactions in the fitted density functions. Furthermore, its implementation is no more difficult than, say, the k-nearest-neighbors method, to which it is similar. Most important, it promises to be quite successful for classification purposes. Aitchison and Aitken tried the binary kernel on a training set of 40 patients known to belong to population A and 37 known to belong to population B. Then it was used to classify 41 unknown patients. All 41 were correctly classified. Anderson (1972), using the same data and a logistic classification procedure, classified only 37 correctly.

## 1.4. MISCELLANEOUS

This section contains a review of some selected techniques relevant to pharmaceutical research. Two of the methods described are clustering algorithms, which seek to identify active clusters of compounds. One is by Harrison and is already well known in the QSAR literature. The other is a new method proposed by one of the authors of this book. The third paper reviewed is a form of AID, and although not presently known in the pharmaceutical literature it promises to be useful. The fourth paper presented describes the key method pioneered by Brown. Although complex, it is nevertheless of considerable methodological interest. The final paper reviews the applications of Gittins's DAIs (dynamic allocation indices) to the problem of discovering lead compounds.

### 1.4.1. Harrison

A procedure based on cluster analysis has been developed by Harrison (1968) and implemented by ICI to discover new lead compounds. It appears to be able to handle moderate quantities of data and some success has been reported in its use.

Like other cluster algorithms, the objective of Harrison's procedure is to identify unusual clusters in a feature space. What is novel in the Harrison approach is the form of the similarity measure used.

In the following it will be assumed that the data set consists of a sequence of vectors each corresponding to a compound, and each of the form $Z = (Y, X_1, \ldots, X_m)$, where Y is the dependent variable and $X_1$ through $X_m$ are the independent variables. All of these variables are assumed to be binary, taking values 1 or 0, which in the case of Y corresponds to the represented compound being active or inactive, and in the case of $X_j$ to feature j being present or absent.

A natural measure of similarity between two compounds is the degree of feature match. However, a weakness of this simple measure is that it does not take into account the relative proportion of compounds possessing the feature in question. For example, if there are two features with respective proportions 0.01 and 0.5, then two compounds sharing the first feature is a relatively more novel event than two compounds sharing the last feature. Consequently, one might be more inclined to consider the first pair more similar than the second pair.

Harrison developed a measure incorporating the observation above by making the following assumption:
1.  Attributes are independently distributed among the compounds.
This is quite a strong assumption and in general will not be true. However, if the procedure developed by Harrison under this assumption leads to useful results, the method will nevertheless be justified.

Under this assumption, if $p_i$ is the probability of selecting at random a compound with attribute i, then the probability of selecting a compound A with a specific feature set is

$$P(A) = \prod_{i=1}^{m} P_i(A)$$

where

$$P_i(A) = \begin{cases} p_i & \text{if A possesses the ith feature} \\ 1 - p_i & \text{otherwise} \end{cases}$$

Moreover, given two compounds A and B, the probability of selecting at random a compound that will have at least as good a match with A on feature i as B has with A is

$$L_i(B \,|\, A) = 1 - Q_i(A)\, \delta_i(A,B)$$

where $Q_i(A) = 1 - P_i(A)$ and $\delta_i(A,B)$ is an indicator function taking the value 1 if and only if A and B match on the ith attribute. Under assumption 1 it then follows that the probability of selecting a compound which will have at least as good a match with A as B for every i is

$$L(B|A) = \prod_{i=1}^{m} L_i(B|A)$$

A measure of similarity between A and B is therefore

$$K(B|A) = -\log L(B|A)$$

$$= -\sum_{i=1}^{m} \log[1 - Q_i(A)\delta_i(A,B)]$$

As may easily be checked, $K(B|A) = K(A|B)$.

One may use any of the available clustering algorithms to discover the clusters in the data set using the similarity measure above. However, the objective here is to find all *significant* clusters of *active* compounds. Harrison approaches this objective by making the following assumption:

2. The proportion of active compounds possessing an attribute is equal to the proportion of inactive compounds possessing that attribute. Hence, if the number of actives among the compounds which are near to a given active compound is higher than is likely under assumption 2 we have statistically significant cluster of active compounds.

Under assumption 2 an estimator for $p_i$ is given by

$$\hat{p}_i = \frac{n_i + 1}{N + 2}$$

where N is the total number of inactives and $n_i$ is the number of inactives having attribute i. It may also be shown that when B is a compound selected at random under assumption 1, the expected value and variance of $K(B|A)$ are given by

$$M_A = -\sum_{i=1}^{m} P_i(A) \log P_i(A)$$

$$V_A = \sum_{i=1}^{m} P_i(A)[1 - P_i(A)][\log P_i(A)]^2$$

Hence the standardized conditional index

$$S(B \mid A) = \frac{K(B \mid A) - M_A}{\sqrt{V}_A}$$

which has approximately a normal distribution with mean 0 and variance 1, may be used for testing the significance of a cluster. The algorithm is as follows:

a. Compute the standardized conditional index of active compound A with each other active compound. Define $d_R$ to be the Rth smallest index.

b. Test for a cluster of compounds around A by examining whether it is significant that out of M -- 1 indexes, where M is the number of active compounds in the data, there are R indexes each of which as a value less than or equal to $d_R$.

c. Procedure b is carried out at a number of significance levels and values of R.

d. Procedures a, b, and c are repeated, replacing compound A with each other active compound in turn.

e. All significant clusters of actives are then listed.

### 1.4.2. Bergman

A new activity clustering algorithm is proposed by one of the present authors (Bergman, 1985). It applies some of the ideas recently presented by Friedman and Rafsky (1979) on minimal spanning trees (MST) to construct a test for the hypothesis that active and inactive compounds are indistinguishable with respect to their features, and to partition the feature space into active and inactive clusters. The strength of this procedure, compared to the algorithm of Harrison, is that it is capable of processing relatively large quantities of data.

It is assumed that the data consist of a set of vectors $(Y, X_1, \ldots, X_m)$, each corresponding to a compound that is known to be active or inactive according to whether Y = 1 or 0 and which possesses features $X_1, \ldots, X_m$. These features may be a realization of a continuous or a binary random variable. In either case we will denote the feature vector $(X_1, \ldots, X_m)$ by Z. It is further assumed that there exists a measure $S(Z_i, Z_j)$ of the degree of similarity between any two feature sets $Z_i$ and $Z_j$ observed for two compounds i and j. How the function S may be defined is a question we shall return to later.

Next we present the necessary graph terminology that will permit us to define a MST. Each feature set $Z_i = (X_{i1}, \ldots, X_{im})$

defines a point in the feature space called a *node*, and a pair $(Z_i, Z_j)$ of such nodes defines an *edge*. An edge may be thought of as a line connecting two nodes in the feature space whose *length* we shall define to be $S(Z_i, Z_j)$. A *path* between two nodes $Z_i$ and $Z_j$ is a sequence of nodes such that the sequence begins with $Z_i$, ends with $Z_j$, and for each intermediate node, all of which are assumed distinct, there is an edge connecting it to the next node in the sequence. A *connected graph* is a set of nodes and edges such that for every pair of nodes there is a path connecting them. A *cycle* is a path beginning and ending with the same node. A *tree* is a connected graph with no cycles; its *length* is the sum of the lengths of the edges in the graph. A *minimal spanning tree* (MST) for a set of nodes is the tree that includes every node and whose total length is a minimum. The MST of the set of nodes reproduced from Friedman and Rafsky in Figure 9 is given in Figure 10.

Under the null hypothesis the probability that the MST will connect an active compound with an inactive compound is proportional to the number of actives and inactives in the data set. Specifically, if there are N compounds and hence N − 1 edges in

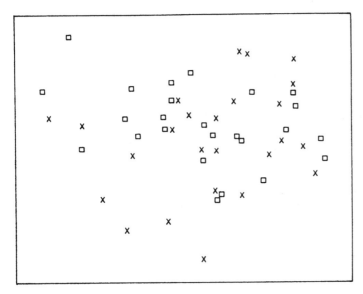

**Figure 9** Actives and inactives. (Figures 9 through 15 on pages 65 through 69 are reprinted with permission of the Institute of Mathematical Statistics.)

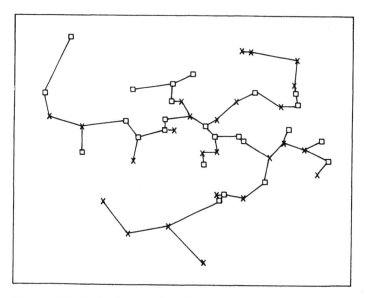

**Figure 10**  Minimal spanning tree.

the MST, the probability that a given edge is such an edge is $mn/N(N - 1)$, where m is the number of actives and n is the number of inactives $(m + n = N)$. A possible test statistic is therefore T, the number of edges in the MST that connect actives to inactives. An alternative statistic is $R = T + 1$, the number of clusters (i.e., connected subgraphs) formed by deleting all such edges from the MST. These ideas are illustrated in Figures 11, 12, and 13. In Figure 11 actives and inactives are plotted in a two-feature Euclidean space. The MST is shown in Figure 12, and the resulting clusters (11 in number) are shown in Figure 13. A similar set of drawings are displayed in Figures 14 and 15.

Using the same arguments as Friedman and Rafsky (1979) used for a different application MSTs, one may derive the expectation and variance of R conditional on C, the number of observed edge pairs that share a common node (in Figure 12, C = 60). One obtains

$$E[R|C] = E[R] = \frac{2m}{N} + 1$$

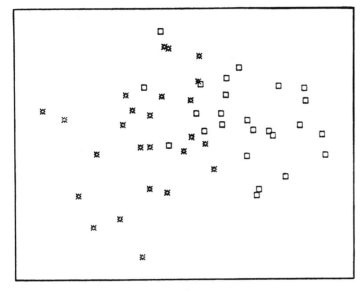

Figure 11  Actives and inactives.

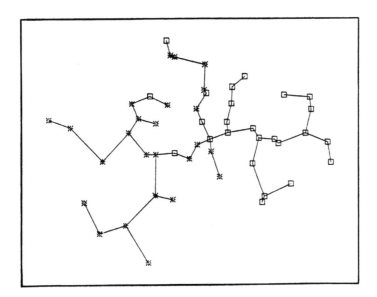

Figure 12  Minimum spanning tree.

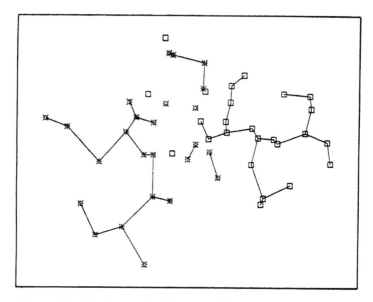

**Figure 13**  Clusters of actives and inactives.

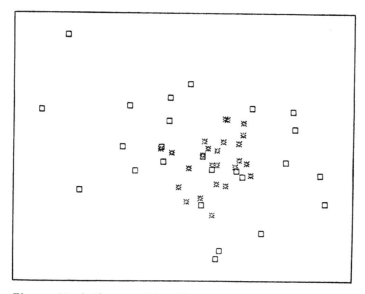

**Figure 14**  Actives and inactives.

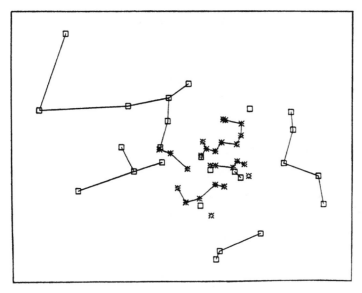

**Figure 15** Clusters of actives and inactives.

$$Var[R\,|\,C] = \frac{2mn}{N(N-1)}\left\{\frac{2mn-N}{N} + \frac{C-N+2}{(N-2)(N-3)}\,[N(N-1)-4mn+2]\right\}$$

The quantity C may be computed by the formula

$$C = \frac{1}{2}\sum_{i=1}^{N} d_i(d_i - 1)$$

where $d_i$ is the number of times node i appears as a member of a pair defining an MST edge. For large samples the distribution of

$$W = \frac{R - E[R]}{(Var[R\,|\,C])^{1/2}}$$

is approximately normal with mean 0 and variance 1. For the graph in Figure 12, W has the value $-4.294$, which is significant at 0.00001. Having established that there is significant information in the feature set for distinguishing actives from inactives the scientist directs his or her attention to the large active clusters to see if their common source of activity can be determined.

Efficient algorithms for computing MSTs are described in the literature. Friedman and Rafsky refer to Whitney (1972), Bentley and Friedman (1975), and Rohlf (1977).

The similarity measure is essentially arbitrary, except that it may not depend on the activity of the compound. One possibility is the number of matched features. Another is the one suggested by Harrison, which is discussed in the preceding section. For binary features another measure that does not depend on any distributional assumptions may be obtained as follows. For given feature sets $Z_i$ and $Z_j$, let $n_{ij}$ be the number of compounds whose features set $Z_k$ satisfies $\min(Z_i, Z_j) \leqslant Z_k \leqslant \max(Z_i, Z_j)$, where $\leqslant$ means that all corresponding components of two vectors satisfy the inequality. Then let $S(Z_i, Z_j) = n_{ij}/N$, where N is the total number of compounds. This measure describes the proportion of the observed compounds whose feature set is as close to $Z_i$ as $Z_j$ for every feature in the feature set. To ensure that $S(Z_i, Z_i) = 0$, the measure above may be modified to be

$$\frac{2n_{ij} - (n_{ii} + n_{jj})}{2N}$$

### 1.4.3. Chi-Squared Automatic Interaction Detection

CHAID (chi-squared AID) is a promising new technique recently developed by Kass (1980) for investigating large quantities of categorical data. It is a refinement of the better known automatic interaction detection (AID) algorithm pioneered by Morgan and Sonquist (1963), and it may be well suited to identifying those common features which contribute to the activity or inactivity of structurally diverse compounds.

To describe how CHAID works, let Z be a typical data vector in a data set consisting of N such vectors. It is assumed that Z is of the form $(Y, X_1, \ldots, X_m)$, where Y is the dependent variable taking any of d possible values, and $X_j$ is an independent variable taking any of $c_j$ possible values. For example, Z may correspond to the measurements taken on a compound, Y being the level of activity (e.g., low, medium, and high) and $X_j$ indicating the presence or absence of a specific molecular fragment. For convenience we may denote the d possible Y values by $1,2,\ldots,d$ and the $c_j$ possible $X_j$ values by $1,2,\ldots,c_j$.

CHAID is a sequential algorithm that begins at stage one by determining for each independent $X_j$ variable the most parsimonious predictor $X_j'$ for Y that can be created by optimally merging the categories of $X_j$. If none of these predictors $X_j'$ is a significant predictor of Y, the algorithm stops. However, if one or more are, the most significant of these is identified and the data set is partitioned in accordance with this predictor's range, thereby creating two or more data subsets. Each of these data subsets now effectively contain $M - 1$ independent variables, and each is separately analyzed and partitioned by the same method used to analyze and partition the original data set. The process stops when no significant partitions can be performed for any of the subsets created by the previous steps. The final result is therefore a dendrogram identifying at each stage the most significant predictor conditional on the partition of the previous stage. Kass gives an example illustrating the type of results that may be obtained.

From the description above it is apparent that the important steps of the Kass algorithm are (1) determining the most parsimonious predictor $X_j'$ for each $X_j$, (2) determining the significance of each $X_j'$, (3) identifying the most significant of these, and (4) establishing a stopping rule for when to cease partitioning. These will be described in detail below.

In defining the method for determining the most parsimonious predictor $X_j'$ for $X_j$, Kass identifies three different types of independent variables: monotonic, free, and floating. These correspond to the permissible ways in which the categories of $X_j$ may be merged. If the values of $X_j$ are part of an ordinal scale, in many cases only adjacent values may be merged to form a new reduced category. Such a variable is called *monotonic*. In contrast, if the categories of $X_j$ are purely nominal, any subset of the original values may form a new collapsed category, and the number of possible $X_j'$ is dramatically increased. Kass calls this a *free* variable. In between these two extremes is the *floating* variable, in which all but one of the range of values belong to an ordinal scale, this exception being purely nominal. Such a variable may arise when a specific code is used for missing data.

The first step for determining $X_j'$ is to form the $c_j \times d$ contingency table depicted below.

$$Y$$

|  | 1 | 2 | 3 |  | d |  |
|---|---|---|---|---|---|---|
| 1 | $n^j_{11}$ | $n^j_{12}$ | $n^j_{13}$ | $\cdots$ | $n^j_{1d}$ | $n^j_{1.}$ |
| 2 | $n^j_{21}$ | $n^j_{22}$ | $n^j_{23}$ | $\cdots$ | $n^j_{2d}$ | $n^j_{2.}$ |
| $\cdot$ | $\cdot$ | $\cdot$ | $\cdot$ | $\cdots$ | $\cdot$ | $\cdot$ |
| $c_j$ | $n^j_{c_j 1}$ | $n^j_{c_j 2}$ | $n^j_{c_j 3}$ | $\cdots$ | $n^j_{c_j d}$ | $n^j_{c_j .}$ |
|  | $n^j_{.1}$ | $n^j_{.2}$ | $n^j_{.3}$ | | $n^j_{.d}$ | N |

$X_j$

Here $n^j_k$ is the number of vectors for which $Y =$ and $X_j = k$, $n^j_{k.}$ is the kth row sum, $n^j$ is the th column sum, and N is the total number of vectors in the data set at this stage. One may note that $n^j$ does not depend on j, as what is being described is only the counts of the Y values, and hence this will be denoted simply by $n$ .

The second step is to construct for each *permissible* pair of categories (k,m) of $X_j$ the $2 \times d$ contingency table

$$Y$$

|  | 1 | 2 | 3 |  | d |  |
|---|---|---|---|---|---|---|
| k | $n^j_{k1}$ | $n^j_{k2}$ | $n^j_{k3}$ | $\cdots$ | $n^j_{k'_d}$ | $n^j_{k.}$ |
| m | $n^j_{m1}$ | $n^j_{m2}$ | $n^j_{m3}$ | $\cdots$ | $n^j_{md}$ | $n^j_{m.}$ |
|  | $n^j_{(k,m)1}$ | $n^j_{(k,m)2}$ | $n^j_{(k,m)3}$ | | $n^j_{(k,m)d}$ | $N^j_{(k,m)}$ |

$X_j$

and test the hypothesis of no association between rows and columns. This may be done by computing the corresponding test statistics

$$T^j_{(k,m)} = \sum_{r=1}^{d} \frac{(n^j_{kr} - e^j_{kr})^2}{e^j_{kr}} + \frac{(n^j_{mr} - e^j_{mr})^2}{e^j_{mr}}$$

where

$$e^j_{kr} = \frac{n^j_{k.} n^j_{(k,m)r}}{N^j_{(k,m)}} \quad \text{and} \quad e^j_{mr} = \frac{(n^j_{m.} n^j_{(k,m)r})}{N^j_{(k,m)}}$$

Under the null hypothesis this statistic has a $\chi^2(d - 1)$ distribution and we define the significance of the table to be the quantity $S(T^j_{(k,m)}) = 1 - F_{d-1}(T^j_{(k,m)})$, where $F_{d-1}(z)$ is the $\chi^2(d - 1)$ distribution function evaluated at z. If this significance is not below some critical level $\alpha$, the two categories are merged. This procedure is now repeated on the new set of (partially) collapsed categories until no new mergers can be performed.

The third step is to test if any of the resulting collapsed categories containing three or more original categories should be split. This is accomplished by finding the most significant binary split into which the collapsed category can be partitioned. If the significance level is below $\alpha$, the indicated split is implemented and the algorithm returns to the second step. Kass states that this third step is necessary to ensure that the procedure does find the (nearly) optimally collapsed contingency table for each independent variable.

When the three-step procedure stops, the nominal significance of the resulting reduced $c'_j \times d$ contingency table may be determined using the statistic

$$T^j_* = \sum_{r=1}^{d} \sum_{k=1}^{c'_j} \frac{(n^j_{kr} - e^j_{kr})^2}{e^j_{kr}}$$

where $e^j_{kr} = (n^j_{k.} n^j_{.r})/N$. The nominal distribution of this statistic is $\chi^2[(c'_j - 1)(d - 1)]$, but as this table was obtained by a selection procedure, the unmodified use of this distribution would be misleading. Kass suggests that an estimate of the true significance level $S(T^j_*)$ may be obtained by multiplying

$1 - F_{(c'_j - 1)(d-1)}(T^j_*)$ by one of the following Bonferroni multipliers, the choice depending on whether $X_j$ is monotonic, free, or floating:

$$B_{monotonic} = \binom{c_j - 1}{c'_j - 1}$$

$$B_{free} = \sum_{i=1}^{c'_j - 1} (-1)^i \frac{(c'_j - i)^{c_j}}{i!(c'_j - i)!}$$

$$B_{float} = \binom{c_j - 2}{c'_j - 2} + c'_j \binom{c_j - 2}{c'_j - 1}$$

Having accomplished the above, the procedure now simply indentifies that $j_0$ for which $S(T^{j_0}_*) = \max_j S(T^j_*)$. Call this maximum $S(T_*)$. If $S(T_*)$ is above some critical value $\beta$, the data set is partitioned into $c'_{j_0}$ subsets such that a vector $Z_n = (Y_n, X_{n1}, \ldots, X_{nm})$ belongs to the ith subset if and only if $X'_{nj_0} = i$.

Kass does not discuss how $\beta$ should be chosen or how $\beta$ may relate to $\alpha$. However, it is clear that since we are selecting the most significant variable out of M variables, the nominal significance of the selected variable for predicting Y considerably overestimates its true predictive power. Consequently, $\beta$ should be chosen to be rather small, perhaps under 0.01.

The computational effort required to achieve the optimal dendrogram does not appear to be excessive. Kass gives as an example a run on 954 vectors consisting of 7 monotonic predictors (with 2 to 9 categories), 11 free predictors (with 6 to 13 categories) and 26 floating predictors (with 10 to 14 categories). On an IBM 370 it took his PL/I program just over 7 minutes to produce 13 final groups.

Kass also suggests that a different version of CHAID based on Fisher's exact test may be derived when each variable is binary. As this is an important case in pharmaceutical research, this possibility will be explored more fully here. Consequently, it is now assumed that each variable takes one of two values (1 or 2), corresponding to active/inactive in the case of the dependent variable, and feature present/not present in the case of the independent variables. Then for each independent variable $X_j$ the following $2 \times 2$ contingency table is constructed:

<br>

|  | Y |  |  |
|---|---|---|---|
|  | 1 | 2 |  |
| $X_j$  1 | $n_{11}^j$ | $n_{12}^j$ | $n_{1.}^j$ |
| 2 | $n_{21}^j$ | $n_{22}^j$ | $n_{2.}^j$ |
|  | $n_{.1}^j$ | $n_{.2}^j$ |  |

<br>

The significance of this table will depend on the type of null hypothesis that is to be tested. Letting $P_1$ and $P_2$ be the conditional probability that the compound will be active given $X_j = 1$ and $X_j = 2$, respectively, we shall consider first the null hypothesis $H_0 : P_1 \geqslant P_2$ against alternative $H_1 : P_1 < P_2$. By standard methods (see Lindgren, 1976, p. 440) it may be shown that conditional on the marginal values $n_{.1}$, $n_{.2}$, and $n_{1.}^j$ the probability of obtaining a table with values

<br>

|  | 1 | 2 |  |
|---|---|---|---|
| 1 | $k$ | $n_{1.}^j - k$ | $n_{1.}^j$ |
| 2 | $n_{.1} - k$ | $n_{.2} - n_{1.}^j + k$ | $N - n_{1.}^j$ |
|  | $n_{.1}$ | $n_{.2}$ |  |

<br>

under $P_1 = P_2$ is given by

$$p_j(k) = \frac{\binom{n_{.1}}{k}\binom{n_{1.}^j - k}{n_{.2}}}{\binom{N}{n_{1.}^j}}$$

We note that this does not depend on the unknown $P_1$. The corresponding test statistic is therefore

$$T_1^j = \sum_{k \geqslant n_{11}^j}^{n_{.1}} p_j(k)$$

A numerically stable method for computing $T_1^j$ that avoids many of the problems involved in computing factorials is the following. Set

$$H_j(k) = \frac{(n_{.1} - k)(n_{1.}^j - k)}{(k + 1)(n_{.2} - n_{1.}^j + k + 1)}$$

and note the relationship

$$p_j(k + 1) = p_j(k)H_j(k)$$

Since $H_j(k)$ is decreasing in k, it follows that $p_j(k)$, $k = 0, \ldots, M$, takes its maximum near

$$k_0 = \min\{M, \max(\hat{k}, 0)\}$$

where $\hat{k}$ is the largest integer k such that

$$k \leqslant \frac{(n_{.1} + 1)n_{1.}^j - (n_{.2} + 1)}{n_{.1} + n_{.2} + 2}$$

Now, set $Q_j(k_0) = 1$ and compute recursively

$$Q_j(i) = \begin{cases} \dfrac{Q(i+1,j)}{H(i)} & \text{for } 0 \leqslant i < k_0 \\[2ex] \dfrac{Q(i-1,j)}{H(i-1)} & \text{for } k_0 < i \leqslant M \end{cases}$$

If $S(j) = \Sigma_{i=0}^{M} Q_j(i)$, one then obtains

$$p_j(i) = \frac{Q_j(i)}{S(j)}$$

for all i.

The quantity $T_1^j$ is considered significant if it is less than some critical value. If there are no significant variables, the process terminates. Otherwise, the most significant variable is identified and the data are partitioned according to its two values.

As an alternative to the test above, one may consider testing the null hypothesis $H_0' : P_1 = P_2$ against the alternative $H_1' : P_1 \neq P_2$. The appropriate test statistic in this case is

$$T_2^j = \min\{T_1^j, 1 - T_1^j + p_j(n_{11}^j)\}$$

Here, too, small values for $T_2^j$ are considered significant.

## 1.4.4.  Brown

A Bayesian approach to the problem of determining the best model for an observed response has been developed by Brown (1970). Although complex, the methodology may be a useful tool for discovering the most active site(s) in a class of congeners. It is assumed that the molecules in the class share a set of major sites, called factors, and that the molecules differ by the substituents occurring at these sites. Each possible substituent at a site is called a state of that factor.

For each molecule in the class a binary response of active or inactive is recorded. A model for this response is a hypothesis about which factors, and which states of these factors, are instrumental in producing activity. For example, assume that there are four factors A, B, C, and D with corresponding sets of states $S_A$, $S_B$, $S_C$, and $S_D$. Let $R_A$ be a subset of $S_A$ and let $\bar{R}_A$ be its complement in $S_A$. Then a possible model is

the hypothesis that activity depends only on factor A and it is
determined by the following rule: a molecule will be active if
and only if its state for factor A belongs to subset $R_A$. It is
easily seen that the number of possible models is quite large.
First there is the question of which set of factors govern the response. For the example there are 15 alternative possibilities:
A, B, C, D, AB, AC, AD, BC, BD, CD, ABC, ACD, ABD, CBD,
and ABCD, which Brown called component keys. Next, for each
component key there is the question of which states correspond
to activity and which do not. For the component key A with $N_A$
states for factor A there are $2^{N_A} - 2$ nontrivial subsets, and
hence $2^{N_A} - 2$ possible models belonging to it. With a component
key containing two or more factors, the possibilities are substantially greater. For example, there are $(2^{N_A} - 2)(2^{N_B} - 2)$
models of the form $\{(a,b) : a \epsilon R_A, b \epsilon R_B\}$ for the component
key AB.

To find the most likely component keys and individual models,
Brown begins by defining a hierarchical prior distribution over
the set of possible models. First, the component keys are
grouped together into classes characterized by the number of factors. Thus component keys A, B, C, and D are the one-factor
keys, while AB, AC, AD, BC, BD, and CD are the two-factor
keys. It is assumed that each of these classes is equally likely.
Then, within a class each component key is considered to be
equally likely. Finally, each model belonging to a component key
is thought to be as likely as any other belonging to that component key.

If we were certain that the biological experiments always re-
vealed the true activity of a compound, it would be a straightforward matter of elimination to determine which models were consistent with the data. In practice one may, however, assume that
there are two types of errors. First, the experiment may show
an active compound to be inactive, and second, it may show an
inactive compound to be active. The error probabilities may be
estimated by performing replicate experiments. Although they
will not be equal in general, we shall assume for simplicity that
they are, and will denote this probability by $\beta(>0)$. Thus even
for the true model one may anticipate a certain degree of discrepancy between the model and the data.

A typical model will be denoted by w and a measure of its discrepancy is $d = d_r + d_s$, where $d_r$ is the number of compounds
predicted active but found inactive and $d_s$ is the number predicted inactive but found active. Under the hypothesis that w is the

true model, the probability of observing d discrepancies in n observations is proportional to

$$(1 - \beta)^{n-d} \beta^d$$

Consequently, by Bayes' theorem, the posterior probability that
w is the true model is also proportional to $(1 - \beta)^{n-d}\beta^d$, which
may be written as

$$(1 - \beta)^n \left(\frac{\beta}{1 - \beta}\right)^d$$

With n fixed and $\beta < 1/2$, this is a monotone decreasing function
of d, and hence models with a small discrepancy will generally be
found to be more likely, given the data, than models with a large
discrepancy.

The discrepancy d for a model may also be decomposed into
components which reveal those aspects of the model contributing
most to the observed discrepancy. For example, suppose that w
is a model within the component key A, and defined by activity
set $R_A$. Let $n_k$ denote the number of observations with state k
for factor A, and $r_k$ and $s_k$ the number of these $n_k$ observations
for which the compound showed activity and inactivity, respectively. If $k \in R_A$, then the discrepancy $q_k$ attributable to state
k is $s_k$, while if $k \notin R_A$, it is $r_k$. It is easily seen that
$d = \sum_{k=1}^{N_A} q_k$. Furthermore, we may note that a certain proportion
of this discrepancy is unavoidable given that we are considering
a model from component key A. Setting $c_k = \min(r_k, s_k)$, $b_k = q_k$
$- c_k$, we obtain $d = \sum_{k=1}^{N_A} c_k + \sum_{k=1}^{N_A} b_k$. The least discrepant
model in component key A has discrepancy $c = \sum_{k=1}^{N_A} c_k$, and
$\sum_{k=1}^{N_A} b_k$ is the amount of discrepancy due specifically to model w.

To go further and obtain the posterior probability that a
component key contains the true model is simply a matter of summing the likelihood of the models within that component key; thus

$$P\left(\begin{array}{l}\text{ith component key} \\ \text{contains true model}\end{array} \Big| \text{data}\right) \propto \sum_{\substack{\text{w belonging to} \\ \text{ith component key}}} P(\text{data}|w)\pi(i)$$

where $\pi(i)$ is the prior probability for each model belonging to the ith component key. Letting $a_{ij}$ denote the number of models within the ith component key which are j discrepant with the data, this is proportional to

$$(1 - \beta)^n \pi(i) \sum_j a_{ij} \left( \frac{\beta}{1 - \beta} \right)^j$$

Setting

$$L_i(\beta) = \sum_j a_{ij} \left( \frac{\beta}{1 - \beta} \right)^j \quad \text{and} \quad L(\beta) = \sum_i \pi(i) L_i(\beta)$$

we then obtain

$$P\left( \begin{matrix} \text{ith component key} \\ \text{contains true model} \end{matrix} \; \middle| \; \text{data} \right) = \frac{L_i(\beta)}{L(\beta)}$$

where $L(\beta)$ acts as the proportionality constant.

To compute $L_i(\beta)$, Brown uses a rather interesting approach which may be exemplified by considering component key A and the associated generating function

$$g_A(x) = \prod_{k=1}^{N_A} (x^{r_k} + x^{s_k}) - (x^r + x^s)$$

where $r = \Sigma_k r_k$ and $s = \Sigma_k s_k$. Expanding $g_A(x)$ into a sum, we obtain

$$g_A(x) = \sum_i x^{\Sigma_{k=1}^{N_A} q_{k_i}}$$

where the sum is over all possible sequences for which $q_{k_i}$ is either $r_k$ or $s_k$. The number of such sequences is $2^{N_A}$, from which we subtract the trivial sequences $(r_1, \ldots, r_{N_A})$ and $(s_1, \ldots, s_{N_A})$, which sum to r and s, respectively. Consequently, the powers of $g_A(x)$ are in one-to-one correspondence with the discrepancies of all models found in component key A. Furthermore,

$$g_A \left( \frac{\beta}{1 - \beta} \right) = L_A(\beta)$$

Setting $t_k = |r_k - s_k|$, we may rewrite $g_A$ as

$$g_A(x) = x^c \prod_{k=1}^{N_A} (1 + x^{t_k}) - (x^s + x^r)$$

Since $\beta/(1 - \beta)$ is usually small, the product above may be approximated by neglecting those terms for which $t_k$ is large. In Table 4 is a set of response data classified according to the component key A. From there we see that

$$g_A(x) = x^{18}(1 + x^0)^2(1 + x^1)^{11}(1 + x^2)^2(1 + x^8)(1 + x^{26})(1 + x^{31})$$

$$- (x^{30} + x^{86})$$

upon collecting together like factors in the expression. This may be approximated by

$$g'_A(x) = 4x^{18}(1 + x)^{11}(1 + x^2)^2$$

after neglecting terms involving powers of x above 22. The prior probability for any model in component key A is $\pi(A)/(2^{18} - 2)$, which may be approximated by $\pi(A)2^{-18}$. Brown carries out a similar set of calculations for component keys B and C (assuming that there is no D component) and finds for his data that the posterior probability for the component key A containing the true model, given that it is contained in a one-factor key model, is 0.75. We note that, according to the data in Table 4, there are four leading models ($R_1 = R$, $R_2 = R \cup \{1\}$, $R_3 = R \cup \{12\}$, $R_4 = R \quad \{1,12\}$, where $R = \{2,3,4,5,6,7,8,10,14,15,17\}$) within component key A which are all least discrepant with respect to A.

By a conditioning argument a similar type of analysis may be performed for keys with two or more factors. Brown also considers the problem of which experiments to perform next in order

**Table 4** Data Table and Posterior Probabilities of Activity Assuming Factor A Is the Component Key (Courtesy Statistician)

| State labels | 4 | 5 | 2, 7 | 3, 8, 10, 14, 15, 17 | 6 | 1, 12 | 9 11, 13, 16 | 18 | |
|---|---|---|---|---|---|---|---|---|---|
| Actives | 43 | 28 | 2 | 1 | 2 | 1 | — | 1 | $86 = \sum_{k=1}^{N_A} r_k$ |
| Probability of activity | 1.00 | 1.00 | 0.98 | 0.89 | 0.89 | 0.50 | 0.11 | 0.00 | |
| Inactives | 12 | 2 | — | — | 1 | 1 | 1 | 9 | $30 = \sum_{k=1}^{N_A} s_k$ |
| $t_k$ | 31 | 26 | 2 | 1 | 1 | 0 | 1 | 8 | |
| $c_k$ | 12 | 2 | 0 | 0 | 1 | 1 | 0 | 1 | $18 = \sum_{k=1}^{N_A} c_k$ |

to gain the maximum information to discriminate between competing models. The details of the analysis of this problem may be found in his article and we content ourselves here with just mentioning the conclusions. First, to discriminate well between two component keys i and j, select an experiment such that component key i predicts the new compound as highly active while component key j predicts it as highly inactive. Second, to discriminate well between models with the same component key, select a compound for testing whose current posterior probability of activity is near 1/2.

## 1.4.5. Dynamic Allocation Indices

### 1.4.5.1. The General Idea

When there are a number of different jobs to be done, projects in which we might invest, or lines of research we might pursue, the question arises of how we should assign priorities so as to minimize costs or maximize rewards. A dynamic allocation index (DAI) is a number associated with any particular alternative, with the property that the optimal policy is to assign priority to the alternative with the largest DAI. These indices typically change as work progresses, so that an optimal policy may well switch back and forth between projects. DAIs with these properties may be defined for a variety of probabilistic models, some of which are relevant to aspects of chemical research. These are reviewed elsewhere (Gittins, 1979, 1982).

One such model has been designed as an aid in the selection of formulations for screening in new-product chemical research. The idea is that within a typical project there are a number of alternative routes representing different possible lines of attack, which vary in difficulty and which may lead toward the solution of the chemical problem for which the project was set up. It is to these routes that DAIs are assigned. The different routes are defined by the different classes of formulations that could be tested. These might be suggested by various chemical hypotheses as to the ways in which the desired result might be achieved, or simply emerge empirically by noting those formulations which have already been found to be reasonably promising. They might, for example, correspond to clusters identified by one of the algorithms described earlier in this chapter.

We suppose that for each formulation a score may be calculated from the test results, the more promising formulations being those with the higher scores. When compounds are being tested to find one that would be suitable for use as a drug, the important thing is the level of therapeutic activity. This is frequently measured by a single number, such as the proportion of diseased animals that recover when treated with the compound, which we may regard as the score for the compound. In this respect, however, pharmaceutical research is the exception rather than the rule. Generally speaking, several different attributes are relevant to the desirability or otherwise of a formulation. However, provided that it is possible to give an order of preference to formulations on the basis of the values taken by the attributes, a single score for each formulation may fortunately still be determined. We shall assume that this has been done, without wishing to suggest that this is always a simple task.

Let T be a *target* score, whose achievement would represent a significant step forward in a project. The choice of T is at the chemist's discretion, but it should be large enough so that a formulation with a score above T is worthy of serious further consideration, and not so large that the project is likely to have to be terminated before such a formulation is found.

A histogram of the scores of the formulations that have so far been tested from a route, showing also the value of T, gives a good indication of the promise of the route, as Figure 16 illustrates. The figure shows the successive histograms for a hypothetical route after 2, 8 and 16 formulations have been tested, respectively. Two possible targets, $T_1$ and $T_2$, are also shown, of which $T_2$ is the more ambitious.

Not much can be said after just two scores have been obtained, although they are sufficiently widely spread to give

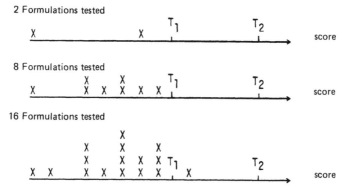

Figure 16 Successive histograms for a route.

grounds for hoping that the targets may both be attainable. When eight formulations have been tested, the picture is clearer. The suggestion that $T_1$ is likely to be reached fairly soon is strengthened, but it is beginning to look as though $T_2$ may not be reached for a long time, unless the chemist hits on some method of finding formulations with higher scores. After testing 16 formulations this picture is confirmed.

The dynamic allocation index for a route is a number that quantifies these impressions obtained from the histogram of scores so far achieved. It is a measure of the current promise of a route, insofar as this is reflected in the scores of the formulations that have already been tested. In many cases the chemist has additional information, perhaps sometimes amounting simply to a hunch, which leads him to believe that a route is either more or less promising than the DAI suggests. When this happens he or she will, quite rightly, take such considerations into account in deciding on which routes to concentrate, and the availability of materials is, of course, also relevant. What the recording of DAIs for the various routes does is to provide an aid to the continual dialogue between the chemist and the experimental data, and to indicate those routes which, on the basis of past results alone, it seems most profitable to pursue.

It is worth noting that the appropriate choice of routes depends on the level of the target. This is illustrated by the route whose history is shown in Figure 16. It is fairly obvious that for the target $T_2$ the DAI for the route must decrease at each successive stage, as it becomes increasingly clear that most of the scores are well below $T_2$. For $T_1$, on the other hand, the DAI may well increase.

This phenomenon is not surprising. It is a reflection of the fact that if what is required is a modest improvement over current performance levels, it is probably best to try modifications of one of the currently used formulations; whereas if the target is a really substantial improvement, it is worth considering completely different and relatively untried routes. It does, however, show the importance of an appropriate choice of target. There may be occasions when it is worthwhile determining DAI values simultaneously for two or more different target values.

A set of DAI tables for routes of this type has been prepared (Gittins and Jones, 1974a), together with suggestions for how to use them. They have been used on an experimental basis by several firms. A computer program that transforms the raw data into a suitable form and avoids the need for tables is nearing completion. The program also provides various other summary statistics, and predicts a range of values describing the number of additional compounds needing to be tested from the route in question before it is likely that the target will be reached.

## 1.4.5.2. Some Details

It is assumed that for any particular route the number of scores greater than x is proportional to $e^{-\lambda x}$ for some $\lambda$, and for all scores above some level I which is less than T. Formulations with scores greater than I are termed interesting formulations. Let q be the fraction of the formulations in the route which are in this category.

In pharmaceutical research there is, as reported by Davies (1962), empirical evidence that the distribution of the activities of compounds is negative exponential. When there is reason to believe that the distribution of activities in a route has a shape that is not negative exponential, a monotone transformation of the scale of scores in order to correct this is desirable. A further linear transformation ensures that I = 0 and T = 100.

The DAI for a route is equal to the posterior probability, given the results from the route to date, that the next formulation tested has a score exceeding T, which we term the current probability of success (CPS), together with an upward correction which is large if the uncertainty associated with that posterior probability is also large. Since our primary interest is in routes for which at most one score over T has so far been noted, it follows that the DAI is something like an extrapolated estimate of the tail area of the distribution of scores in the route, based on the current histogram of scores. This means that the DAI is subject to considerable error if there is any appreciable departure from a negative exponential distribution of scores, and any such error will be particularly great if the extrapolated distribution is fitted to those parts of the histogram for which the scores are well below T. For this reason formulations with scores less than I are not used in the estimation of $\lambda$.

Before experimentation commences on a route, all the possible values for q and $\lambda$ are assumed to be equally likely. More precisely, it is assumed that q and $\lambda$ have prior distributions which are independent, q having a uniform distribution on $(0,1)$, and $\lambda$ having an improper prior distribution on $(0,\infty)$ which is proportional to $\lambda^{-1}$, so that log $\lambda$ is uniformly distributed on $(-\infty,\infty)$. It follows that after m + n formulations have been tested, m of them having proved to be uninteresting and the other n scoring $x_1, x_2, \ldots, x_n$, then q has a posterior beta distribution with the parameters m and n, and $\lambda$ a posterior gamma distribution with the parameters n and $\Sigma x_i$. Like the prior distributions, these posterior distributions are independent. The posterior probability that the next formulation tested will be interesting is $(n + 1)/(n + m + 2)$, and the current probability of success is $(n + 1)(\Sigma x_i)^n / [(n + m + 2)(100 + \Sigma x_i)^n]$.

Strictly speaking, the assumptions just described imply that the probability of a compound achieving a particular score depends only on the route to which it belongs. This overlooks the fact that the degree of similarity between compounds typically varies within a route, and compounds of similar structure are more likely to achieve similar scores. This difficulty may be reduced by including in the construction of the DAI only a representative selection of those compounds that form a group with closely similar structures within a route that includes other more diverse compounds. Suggestions on how to make this selection, and details of the definition and method of calculation of the DAIs, are given by Gittins and Jones.

## 1.4.5.3. An Example

In Figure 17 the results of tests on formulations drawn from two routes in a research program designed to produce a herbicide are summarized. The raw data for each formulation consisted of assessments of the severity of the effect when the formulation was applied to a particular plant species in a particular way and at a given dosage level. In all, forty observations were available for each formulation for different combinations of these factors. The target was represented by an existing herbicide, the aim being to find a formulation which is at least as toxic to the relevant plant species. Similar raw data was available for the target herbicide. Scores for each formulation were obtained by first taking an appropriate weighted average of the differences between the forty observations for target herbicide and formulation respectively, and then adding a constant so that the lowest observed score was zero.

From these scores the DAI program selected the indicated values of I so as to achieve a balance between the aims of enough scores above I to give reasonable estimates and not so many as to produce an avoidable risk of extrapolation errors. The program also selected the indicated transformations to achieve a closer fit to negative exponential distributions. It then calculated CPS and DAI for each route assuming ignorance prior distributions for the parameters.

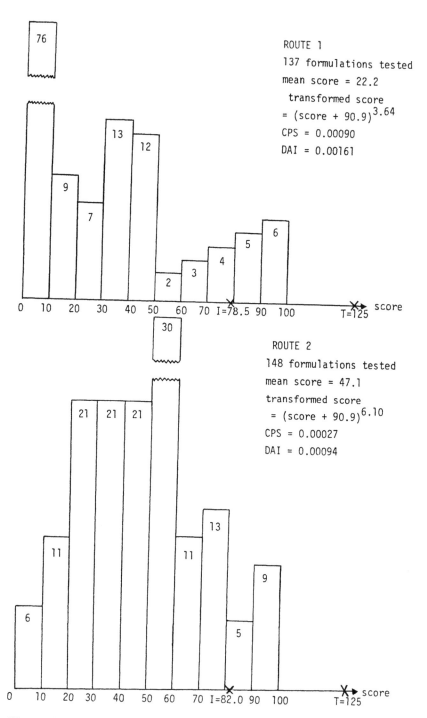

ROUTE 1
137 formulations tested
mean score = 22.2
 transformed score
= $(\text{score} + 90.9)^{3.64}$
CPS = 0.00090
DAI = 0.00161

ROUTE 2
148 formulations tested
mean score = 47.1
transformed score
= $(\text{score} + 90.9)^{6.10}$
CPS = 0.00027
DAI = 0.00094

Figure 17  DAI Analysis for two routes in herbicide research.

From the CPS values it is apparent that for both routes the number of further formulations needing to be tested before the target is reached may well run into thousands. This is a plausible inference from the histograms themselves.

It is interesting to note that although route 2 has produced more formulations with scores near the target, it is route 1 that has the higher CPS and DAI. This is because of the higher power used in the transformation for route 2, which in turn is because the raw data for route 1 are more nearly negative exponential than for route 2. Before attaching any significance to this comparison we must consider whether there are any artificial features about the two score distributions. In this case it turned out that for both routes the relatively large number of formulations with scores near the maximum is because several formulations all with very similar molecular structure were tested. The next step was to remove some of these from the analysis, as mentioned in the previous section.

## REFERENCES

Abe, H., and Jurs, P. C. (1975), "Automated chemical structure analysis of organic molecules with a molecular structure generator and pattern recognition techniques," *Anal. Chem.*, Vol. 47, pp. 1829-1835.

Adamson, G. W., and Bawden, D. (1975), "A method of structure-activity correlation using WLN," *J. Chem. Inf. Comput. Sci.*, Vol. 15, pp. 215-220.

Adamson, G. W., and Bawden, D. (1976), "An empirical method of structure-activity correlation for polysubstituted cyclic compounds using WLN," *J. Chem. Inf. Comput. Sci.*, Vol. 16, pp. 161-165.

Adamson, G. W., and Bush, J. A. (1973), "A method for the automatic classification of chemical structures," *Inf. Stor. Ret.*, Vol. 9, pp. 561-568.

Adamson, G. W., and Bush, J. A. (1974), "Method for relating the structure and properties of chemical compounds," *Nature*, Vol. 248, pp. 406-408.

Adamson, G. W., and Bush, J. A. (1975), "A comparison of the performance of some similarity and dissimilarity measures in the automatic classification of chemical structures," *J. Chem. Inf. Comput. Sci.*, Vol. 15, pp. 55-58.

Adamson, G. W., and Bush, J. A. (1976), "Evaluation of an empirical SAR for property prediction in a structurally diverse group of local anaesthetics," *J. Chem. Soc. Perkin 1*, pp. 168-172.

Adamson, G. W., Lynch, M. F., and Town, W. G. (1971), "Analysis of structural characteristics of chemical compounds in a large computer-based file: Pt. 2. Atom-centred fragments," *J. Chem. Soc.* Vol. (C), pp. 3702-3706.

Adamson, G. W., Lambourne, D. R., and Lynch, M. F. (1972), "Analysis of structural characteristics of chemical compounds in a large computer-based file: Pt. 3. Statistical association of fragment incidence," *J. Chem. Soc. Perkin 1*, pp. 2428-2433.

Adamson, G. W., Cowell, J., Lynch, M. F., McClure, A. H., Town, W. G., and Yapp, A. (1973a), "Strategic considerations in the design of a screening system for substructure searches of chemical structure files," *J. Chem. Doc.*, Vol. 13, pp. 153-157.

Adamson, G. W., Cowell, J., Lynch, M. F., Town, W. G., and Yapp, A. M. (1973b), "Analysis of structural characteristics of chemical compounds in a large computer-based file: Pt. 4. Cyclic fragments," *J. Chem. Soc. Perkin 1*, pp. 863-865.

Adamson, G. W., Greasey, S. E., Eakins, J. P., and Lynch, M. F. (1973c), "Analysis of structural characteristics of chemical compounds in a large computer-based file: Pt. 5. More detailed cyclic fragments," *J. Chem. Soc. Perkin 1*, pp. 2071-2076.

Adamson, G. W., Greasey, S. E., and Lynch, M. F. (1973d), "Analysis of structural characteristics of chemical compounds in the common data base," *J. Chem. Doc.*, Vol. 13, pp. 158-162.

Adamson, G. W., Bush, J. A., McClure, A. H., and Lynch, M. F. (1974), "An evaluation of a substructure search system based on bond-centered fragments," *J. Chem. Doc.*, Vol. 14, pp. 44-48.

Aitchison, J., and Aitken, C. G. (1976), "Multivariate binary discrimination by the kernel method," *Biometrika*, Vol. 63, No. 3, pp. 413-420.

Aitchison, J., and Begg, B. (1976), "Statistical diagnosis when basic cases are not classified with certainty," *Biometrika*, Vol. 63, No. 1, pp. 1-12.

Anderson, J. A. (1972), "Separate sample logistic discrimination," *Biometrika*, Vol. 59, pp. 19-35.

Anderson, J. A. (1974), "Diagnosis by logistic discriminant function: further practical problems and results," *Appl. Statist.*, Vol. 23, No. 3, pp. 397-404.

Anderson, J. A. (1975), Quadratic logistic descrimination, *Biometrika*, Vol. 62, No. 1, pp. 149-154.

Anderson, J. A., Whaley, K., Williamson, J., and Buchanan, W. W. (1972), "A statistical aid to the diagnosis of keratoconjunctivitis sicca," *Quart. J. Med.*, Vol. 41, pp. 175-189.

Andrews, H. C. (1972), *Introduction to Mathematical Techniques in Pattern Recognition*, Wiley-Interscience, New York.

Ash, J. E., and Hyde, E. (1975), *Chemical Information Systems*, Ellis Horwood, Chichester, England.

Avidon, V. V., and Leksina, L. A. (1974), "A descriptor language for analyzing the similarity of the chemical structure of organic compounds," *Autom. Doc. Math. Linguist.*, Vol. 8, pp. 61-65.

Balaban, A. T., et al. "Steric fit in QSAR," *Lecture Notes in Chemistry*, No. 15, G. Berthier, ed., Springer-Verlag, New York.

Bawden, D. (1978), "Substructural analysis techniques for structure-property correlation within computerized chemical information systems," Ph.D. thesis, University of Sheffield.

Beale, E. M. L., Kendall, M. G., and Mann, D. W. (1967), "The discarding of variables in multivariate analysis," *Biometrika*, Vol. 54, pp. 357-366.

Bender, C. F., and Kowalski, B. R. (1974), "Multiclass linear classifier for spectral interpretation (pattern recognition)," *Anal. Chem.*, Vol. 46, pp. 294-296.

Bentley, J. L., and Friedman, J. H. (1975), "Fast algorithms for constructing minimal spanning trees in coordinate spaces," *Stamford Linear Accelerates Report (SLAC) PUB-1665*.

Bergman, S. W. (1985), "A discriminant clustering algorithm with pharmaceutical applications." In preparation.

Bishop, Y. M., Fienberg, S. E., and Holland, P. W. (1975), *Discrete Multivariable Analysis: Theory and Practice*, MIT Press, Cambridge, Mass.

Box, G. E., and Draper, N. R. (1969), *Evolutionary Operation*, Wiley, New York.

Box, and Hunter, (1957), "Multi-factor experimental design for exploring response surfaces," *Ann. Math. Statist.*, Vol. 28, pp. 195-241.

Boyce, A. J. (1969), "Mapping diversity: a comparative study of some numerical methods," in *Numerical Taxonomy, Proceedings of the Colloquium in Numerical Taxonomy*, held at the University of St. Andrews, A. J. Cole, ed., Academic Press, London.

Brock, N., and Schneider, B. (1961), "Pharmacological screenings of drugs by means of the therapeutic index," in *Quantitative Methods in Pharmacology*, 2nd ed., H. DeJonge, ed., North-Holland, Amsterdam.

Brown, P. J. (1971), "Keys to correlate biological activity with molecular structure of chemical compounds," *Statistician*, Vol. 20, pp. 3—49.

Brown, P. J. (1973), "Aspects of design for binary key models," *Biometrika*, Vol. 60, No. 2, pp. 309-318.

Brown, P. J. (1976), "Remarks on some statistical methods for medical diagnosis," *J. R. Statist. Soc. A (General)*, Vol. 139, Part 1, pp. 104-107.

Brown, P. J. (1977), "Functions for selecting tests in diagnostic key construction," *Biometrika*, Vol. 64, No. 3, pp. 589-596.

Burger, A. (1960), "Relation of chemical structure and biological activity," in *Medicinal Chemistry*, 2nd ed., A. Burger, ed., Interscience, New York.

Burger, A. (1980), *Medicinal Chemistry*, 4th ed., Wiley-Interscience, New York.

Bush, J. A. (1977), "Automatic classification and chemical structure-activity correlation," Ph.D. thesis, University of Sheffield.

Cammarata, A. (1970), "Quantitative SAR," *Ann. Rep. Med. Chem.* (Sec. 4), pp. 245-253.

Cammarata, A., and Menon, G. K. (1976), "Pattern recognition. Classification of therapeutic agents according to pharmacophores," *J. Med. Chem.*, Vol. 19, pp. 739-748.

Chang, C. L., and Lee, R. C. T. (1973), *IEEE Trans. Syst. Man. Cybern.*, Vol. SMC-3, p. 197.

Chou, J. T., and Jurs, P. C. (1979a), "Computer assisted structure-activity studies of chemical carcinogens. An N-nitroso compound data set," *J. Med. Chem.*, Vol. 22, pp. 792-797.

Chou, J. T., and Jurs, P. C. (1979b), "Computer-assisted computation of partition coefficients from molecular structures using fragment constants," *J. Chem. Inf. Comput. Sci.*, Vol. 19, No. 3., pp. 172-178.

Chu, K. C. (1974), "Application of artificial intelligence to chemistry. Use of pattern recognition and cluster analysis to determine pharmacological activity of some organic compounds," *Anal. Chem.*, Vol. 46, pp. 1181-1187.

Chu, K. C. (1980), "The quantitative analysis of structure-activity relationships," in *Medicinal Chemistry*, A. Burger, ed., Vol. I, Wiley-Interscience, New York, pp. 393-418.

Chu, K. C., Feldman, R. J., Shapiro, M. B., Hazard, G. F., and Gerar, R. I. (1974), "Pattern recognition and structure-activity relationship studies. Computer-assisted prediction of antitumor activity in structurally diverse drugs in an experimental mouse brain tumor system," *J. Med. Chem.*, Vol. 18, No. 6, pp. 539-545.

Clark, H. A., and Jurs, P. C. (1975), "Qualitative determination of petroleum sample type from gas chromatograms using pattern recognition techniques," *Anal. Chem.*, Vol. 47, pp. 374-378.

Cover, T. M., and Hart, P. E. (1967), "Nearest neighbor pattern classification," *IEEE Trans. Inf. Theory*, Vol. IT-13, pp. 21-27.

Craig, P. N. (1971a), "Comparison of the Hansch and Free-Wilson approaches to structure-activity correlation," *Biological Correlations—The Hansch approach*, Advances in Chemistry Series Vol. 114, ed. W. van Valkeburg, Amer. Chem. Soc., pp. 115-129.

Craig, P. N. (1971b), "Interdependence between physical parameters and selection of substituent groups for correlation studies," *J. Med. Chem.*, Vol. 14, pp. 680-684.

Craig, P. N. (1975), "Structure/Property Correlations," in *Chemical Information Systems*, J. E. Ash, and E. Hyde, eds., Ellis Horwood, New York.

Craig, P. N. (1978), *Hansch Analysis, Free-Wilson Analysis, and Substructure Analysis*, DHEW Publ., FDA 798-1046, Struct. Correl. Carinog. Mutagen.

Cramer, R. D. (1973), "Substructural analysis, a novel approach to the problem of drug design," *J. Med. Chem.*, Vol. 17, No. 5, pp. 533-535.

Crippen, G. (1979), "Distance geometry approach to rationalizing binding data," *J. Med. Chem.*, Vol. 22, No. 8, pp. 988-997.

Crowe, J. E., Lynch, M. F., and Town, W. G. (1970), "Analysis of structural characteristics of chemical compounds in a large computer-based file: Pt. 1. Non-cyclic fragments," *J. Chem. Soc.*, Vol. (C), pp. 990-996.

Crum-Brown, A., and Fraser, T. (1869), *Trans. R. Soc. Edinb.*, Vol. 25, No. 151, p. 693.

Darvas, F. (1974), "Application of the sequential simplex method in designing drug analogs," *J. Med. Chem.*, Vol. 17, No. 8, pp. 799-804.

Das Gupta, S. (1973), "Theory and methods in classification: a review," in Discriminant Analysis and Prediction, T. Cacoullos, pp. 77-137.

Davies, O. L. (1962), "Some statistical considerations in the selection of projects for research in the pharmaceutical industry," *Appl. Statist.*, Vol. 11, pp. 170-183.

Day, N. E., and Kerridge, D. F. (1967), "A general maximum
likelihood discriminant," *Biometrics*, Vol. 23, pp. 313-323.

Dierdorf, D. S., and Kowalski, B. R. (1974), "Three dimensional
molecular structure-biological activity correlation by pattern
recognition," *NTIS Report AD 785 863.*

Dirren, H., Robinson, A. B., and Pauling, L. (1975), *Clin.
Chem.*, Vol. 21, p. 1970.

Dixon, W. J. (1967), *BMD; Biomedical Computer Programs,*
University of California Press, Berkeley, Calif.

Draper, N., and Smith, H. (1966), *Applied Regression Analysis,*
Wiley, New York.

Dudewicz, E. (1980), "Ranking (ordering) and selection: an
overview of how to select the best," *Technometrics*, Vol. 22,
No. 1.

Duewer, D. L., Kowalski, B. R., and Schatzki, T. F. (1975),
"Source identification of oil spills by pattern recognition
analysis of natural elemental composition," *Anal. Chem.*,
Vol. 47, pp. 1573-1583.

Dunn, W. J., and Wald, S. (1980), "Structure activity analyzed
by pattern recognition: the asymmetric case," *J. Med. Chem.*,
Vol. 23, pp. 595-599.

Dunn, W. J., Greenberg, M. J., and Cammarata, A. (1976), "Use
of cluster analysis in the development of SAR for antitumor
triazines," *J. Med. Chem.*, Vol. 19, pp. 1299-1301.

Efroymson, M. A. (1960), "Multiple regression analysis," in
*Mathematical Methods for Digital Computers*, A. Ralston and
H. S. Wilf, eds., Vol. 1, pp. 191-203.

Everitt, B. (1974), *Cluster Analysis*, Heinemann, London.

Fix, E., and Hodges, J. L. (1951), "Discriminatory analysis,
nonparametric discrimination," US Air Force School of Avia-
tion Medicine, Randolf Field, Tex, Prospect 21-49-004,
Report 4, Contract AF41 (128)-31.

Forgey, E. W. (1965), "Cluster analysis of multivariate data:
efficiency versus interpretability of classification," *Biometrics*, Vol. 21, pp. 768-769.

Free, S. M., and Wilson, J. W. (1964), "A mathematical contribution to structure-activity studies," *J. Med. Chem.*, Vol. 7,
No. 4, pp. 395-399.

Friedman, J. H., and Rafsky, L. C. (1979), "Multivariable generalization of the Wald-Wolfowitz and Smirnov two sample
tests," *Ann. Statist.*, Vol. 7, No. 4, pp. 697-717.

Fu, K. S. (1968), *Sequential Methods in Pattern Recognition and Machine Learning*, Academic Press, New York.

Gittins, J. C. (1971), "An index for sequential project selection," *R&D Manage.*, Vol. 1, No. 3, pp. 37-140.

Gittins, J. C. (1979), "Bandit processes and dynamic allocation indices," *J. R. Statist. Soc. B*, Vol. 14.

Gittins, J. C. (1982), "Forwards induction and dynamic allocation indices," *Proc. NATO Conference on Deterministic and Stochastic Scheduling, Durham 1981*, Reidel, Amsterdam.

Gittins, J. C., and Jones, D. M. (1974a), "A Dynamic Allocation Index for New-Product Chemical Research," Department of Engineering Technical Report, University of Cambridge, Cambridge.

Gittins, J. C., and Jones, D. M. (1974b), "A dynamic allocation index for the sequential design of experiments," *Proc. Eur. Meet. Statist.*, Hungarian Academy of Sciences, Budapest; 1972.

Goldstein, M., and Dillon, W. R. (1978), *Discrete Discriminant Analysis*, Wiley, New York.

Habbema, J. D. F., Hermans, J., and Van Den Brock, K. (1974), "A stepwise discriminant analysis program using density estimation," in *Compstat 1974*, G. Brickmann, ed., Physica Verlag, Vienna; pp. 101-110.

Haberman, S. (1974), *The Analysis of Frequency Data*, University of Chicago Press, Chicago.

Hansch, C. (1967), "Physicochemical parameters in drug design," *Annu. Rev. Med. Chem.*, pp. 348-357.

Hansch, C. (1969), "A quantitative approach to biochemical structure-activity relationships," *Acc. Chem. Res.*, Vol. 2, pp. 232-239.

Hansch, C. (1971), "Quantitative SAR in drug design," *Drug Design 1*, E. J. Ariens, ed., Medicinal Chemistry Series, Vol. 11, Academic Press, New York, pp. 271-342.

Hansch, C. (1973), "Aromatic substituent constants for structure-activity correlations," *J. Med. Chem.*, Vol. 16, pp. 1207-1216.

Hansch, C. (1975), "On the structure of medicinal chemistry," *J. Med. Chem.*, Vol. 19, pp. 1-6.

Hansch, C., and Anderson, S. M. (1967), "The structure-activity relationships in barbiturates and its similarity to that in other narcotics," *J. Med. Chem.*, Vol. 10, pp. 745-753.

Hansch, C., Unger, S. H., and Forsythe, A. B. (1973), "Strategy in drug design. Cluster Analysis as an aid in the selection of substituents," *J. Med. Chem.*, Vol. 16, pp. 1217-1222.

Hansch, C., Rockwell, S. D., Jow, P. Y. C., Leo, A., and Steller, E. E. (1976), "Substituent constants for correlation analysis," *J. Med. Chem.*, Vol. 20, No. 2, pp. 304-308.

Harrison, P. J. (1968), "A method of cluster analysis and some applications," *J. R. Statist. Soc. C*, Vol. 17.

Hartigan, J. A. (1975), *Clustering Algorithms*, Wiley, New York.

Hawkins, D. M. (1973), "On the investigations of alternative regressions by principal component analysis," *Appl. Statist.*, Vol. 22, pp. 275-286.

Heller, S. R. (1974), "Computer Techniques for Interpreting Mass Spectrometry Data," in *Computer Representation and Manipulation of Chemical Information*, W. T. Wipke, ed., Wiley, New York.

Henry, D., and Block, J. (1979a), "Classification of drugs by discriminant analysis using fragment molecular connectivity values," *J. Med. Chem.*, Vol. 22, No. 5, pp. 465-472.

Henry, D. R., and Block, J. H. (1979b), "Steroid classification by discriminant analysis using fragment molecular connectivity," *Eur. J. Med. Chem.-Chim. Ther.*, Vol. 15, No. 2, pp. 133-138.

Hermans, J., and Habbema, J. D. F. (1975), "Comparison of five methods to estimate posterior probabilities," *EDV Med. Biol.*, Vol. 6.

Hiller, S. A. (1973), "The perceptron approach," *Comp. Biomed. Res.*, Vol. 6, p. 411.

Ho, Y.-C., and Kashyap, R. L. (1965), *IEEE Trans. Electron. Comput.*, Vol. EC-14, No. 5, pp. 682-688.

Hodes, L., Hazard, G. F., Geran, R. I., and Richman, S. (1977), "A statistical-heuristic method for automated selection of drugs for screening," *J. Med. Chem.*, Vol. 20, pp. 469-475.

Isenhour, T. D., and Jurs, P. (1971), Chemical applications of machine intelligence, *Anal. Chem.*, Vol. 48, p. 20A-35A.

Isenhour, T. L., and Jurs, P. C. (1973), "Learning machines," in *Computer Fundamentals for Chemists*, J. S. Mattson, H. B. Mark, and H. C. MacDonald, eds., Marcel Dekker, New York, pp. 285-331.

Isenhour, T. L., Kowalski, B. R., and Jurs, P. C. (1974), "Applications of pattern recognition to chemistry," *CRC Crit. Rev. Anal. Chem.*, Vol. 4, pp. 1-39.

Jardine, N., and Van Rijsbergen, C. J. (1971), "The use of hierarchic clustering in information retrieval," *Inf. Stor. Ret.*, Vol. 7, pp. 217-240.

Jeffers, J. N. R. (1981), "Investigations of alternative regressions: some practical examples," *Statistician*, Vol. 30, No. 2, pp. 79-88.

Jurs, P. C. (1970), "Mass spectral feature selection and struc-
  tural correlations using computerized learning machines,"
  *Anal. Chem.*, Vol. 42, pp. 1633-1638.
Jurs, P. C. (1974), "Chemical data interpretation using pattern
  recognition techniques," in *Computer Representation and
  Manipulation of Chemical Information*, W. T. Wipke, ed.,
  Wiley, New York.
Jurs, P. C. (1978), *Pattern Recognition Methods*, DHEW Publ.
  FDA 79-1046, Struct. Correl. Carcinog. Mutagen.
Jurs, P. C., and Isenhour, T. L. (1975), *Chemical Applications
  of Pattern Recognition*, Wiley, New York.
Jurs, P. C., Kowalski, B. R., and Isenhour, T. L. (1969a),
  "Computerized learning machines applied to chemical prob-
  lems. Molecular formula determination from low resolution
  mass spectrometry," *Anal. Chem.*, Vol. 41, pp. 21-27.
Jurs, P. C., Kowalski, B. R., Isenhour, T. L., and Reilley,
  C. N. (1969b), "An investigation of combined patterns from
  diverse analytical data using computerized learning machines,"
  *Anal. Chem.*, Vol. 41, pp. 1949-1953.
Jurs, P. C., Kowalski, B. R., Isenhour, T. L., and Reilley,
  C. N. (1969c), "Computerized learning machines applied to
  chemical problems. Investigation of convergence rate and
  predictive ability of adaptive binary pattern classifiers,"
  *Anal. Chem.*, Vol. 41, pp. 690-694.
Jurs, P. C., Kowalski, B. R., Isenhour, T. L., and Reilley,
  C. N. (1970), "Computerized learning machines applied to
  chemical problems. Molecular structure parameters from low
  resolution mass spectrometry," *Anal. Chem.*, Vol. 42, pp.
  1387-1394.
Justice, J. B., and Isenhour, T. L. (1974), "Information content of mass spectra as determined by pattern recognition
  methods," *Anal. Chem.*, Vol. 46, pp. 223-226.
Kaminuma, T., Tatekawa, T., and Watanabe, S. (1969), "Reduc-
  tion of clustering problem to pattern recognition," *Pattern
  Recogn.*, Vol. 1, pp. 195-205.
Kanal, L. (1974), "Patterns in pattern recognition 1968-1974,"
  *IEEE Trans. Inf. Theory*, Vol. IT-20, No. 6, pp. 697-722.
Kass, G. V. (1980), "An exploratory technique for investigating
  large quantities of categorical data," *Appl. Statist.*, Vol. 29,
  No. 2, pp. 119-127.
Kendall, M. (1975), *Multivariate Analysis*, Griffin, High Wycombe,
  England.
Kier, L. B., and Hall, L. H. (1976), "The nature of structure-
  activity relationships and their relation to molecular connec-

tivity," *Eur. J. Med. Chem.-Chim. Ther.*, Vol. 12, No. 4, pp. 307-308.

Kirschner, G., and Kowalski, B. (1979), "The application of pattern recognition to drug design," *Med. Chem. (Academic)*, Vol. 11, No. 8, pp. 73-131.

Koskinen, J. R., and Kowalski, B. R. (1974), "Structure-activity correlations for organic molecules by pattern recognition," *NTIS Report AD-785 913.*

Koskinen, J. R., and Kowalski, B. R. (1975), "Interactive pattern recognition in the chemical laboratory," *J. Chem. Inf. Comput. Sci.*, Vol. 15, pp. 119-123.

Kowalski, B. R. (1974), "Pattern recognition in chemical research," *Computers in Chemical and Biochemical Research*, C. E. Klopfenstein, and C. L. Wilkins, eds., Vol. 2, Academic Press, New York.

Kowalski, B. R. (1975), "Meanverment analysis by pattern recognition," *Anal. Chem.*, Vol. 47, p. 1152A-1162A.

Kowalski, B. R., and Bender, C. F. (1972a), "Pattern recognition. A powerful approach to interpreting chemical data," *J. Am. Chem. Soc.*, Vol. 94, pp. 5632-5639.

Kowalski, B. R., and Bender, C. F. (1972b), "The K-nearest neighbour classification rule (pattern recognition) applied to nuclear magnetic resonance spectral interpretation, *Anal. Chem.*, Vol. 44, pp. 1405-1411.

Kowalski, B. R., and Bender, C. F. (1973), "Pattern recognition: II. Linear and nonlinear methods for displaying chemical data," *J. Am. Chem. Soc.*, Vol. 95, pp. 686-693.

Kowalski, B. R., and Bender, C. F. (1974), "The application of pattern recognition to screening prospective anticancer drugs. Adenocarcinoma 755 biological activity test," *J. Am. Chem. Soc.*, Vol. 96, pp. 916-918.

Kowalski, B. R., Jurs, P. C., Isenhour, T. L., and Reilley, C. N. (1969), "Computerized learning machines applied to chemical problems. Multicategory pattern classification by least squares," *Anal. Chem.*, Vol. 41, pp. 695-700.

Kowalski, B. R., Schatzki, T. F., and Striss, F. H. (1972), "Classification of archaelogical artifacts by applying pattern recognition to trace element data," *Anal. Chem.*, Vol. 44, pp. 2176-2180.

Kubinyi, H. (1976), QSAR relationships, 2. A mixed approach based on Hansch and Free-Wilson analysis, *J. Med. Chem.*, Vol. 19, p. 587.

Kuo, K.-S., and Jurs, P. C. (1973), "Semi-quantitative determination of chlorine dosages for water treatment using pattern recognition techniques," *J. Am. Water Works Assoc.*, Vol. 65, p. 623-626.

Leo, A., Hansch, C., and Church, C. (1969), "Comparisons of parameters currently used in the study of SAR," *J. Med. Chem.*, Vol. 12, pp. 766-771.

Liddell, R. W., and Jurs, P. C. (1974), "Interpretation of infrared spectra using pattern recognition techniques," *Anal. Chem.*, Vol. 46, pp. 2126-2130.

Lindgren, B. W. (1976), *Statistical Theory*, Macmillan, New York.

Martin, Y. C. (1978), *Quantitative Drug Design*, Medicinal Research Series, Vol. 8, G. L. Grunewald, ed., Marcel Dekker, New York.

Martin, Y. C., and Panas, H. N. (1979), "Mathematical considerations in series design," *J. Med. Chem.*, Vol. 22, No. 7, pp. 784-791.

Martin, Y. C., Holland, J. B., Jarboe, C. H., and Plotnikoff, N. (1974), "Discriminant analysis of the relationship between physical properties and the inhibition of monoamine oxidase by aminotetralins and aminoindans," *J. Med. Chem.*, Vol. 17, pp. 409-413.

Massart, D. L., and Kaufman, L. (1975), Optimisation of flow schemes for ion exchange separations by dynamic programming," *Anal. Chem.*, Vol. 47, p. 1244A.

Massy, W. F. (1965), "Principal component regression in exploratory statistical research," *J. Am. Statist. Assoc.*, Vol. 60, pp. 234-256.

Mathews, R. J. (1975), "A comment on the structure-activity correlations obtained using pattern recognition methods," *J. Am. Chem. Soc.*, Vol. 97, pp. 935-936.

Maxwell, A. E. (1977), *Multivariate Analysis in Behavioral Research*, Chapman & Hall, London.

McConnell, L., Rhodes, G., and Watson, U. (1979), "Application of pattern recognition and feature extraction techniques to volatile constituent metabolic profiles obtained by capillary gas chromatography," *J. Chromatogr. Biomed. Appl.*, Vol. 162, pp. 495-506.

McGill, J. R., and Kowalski, B. R. (1978), "Classification of mass spectra via pattern recognition," *J. Chem. Inf. Comput. Sci.*, Vol. 18, pp. 52-55.

Meisel, W. S. (1972), *Computer-Oriented Approaches to Pattern Recognition*, Academic Press, New York.

Mendel, J. M., and Fu, K. S. (1970), *Adaptive Learning and Pattern Recognition Systems*, Academic Press, New York.

Minsky, M., and Papert, S. (1969), *Perceptions*, MIT Press, Cambridge, Mass.

Morgan, J. A., and Sonquist, J. N. (1963), "Some Results from a non-symmetric branching process that looks for interaction effects." *Proc. Soc. SPATS Sec. ASA*, pp. 40-53.

Moriguchi, I., Komatsu, K., and Matsushita, Y. (1979), "Adaptive least-squares method applied to structure-activity correlation of hypotensive $N$-alkyl-$N''$-cyano-$N'$-pyridylguaridines," *J. Med. Chem.*, Vol. 23, pp. 20-26.

Nilsson, N. J. (1965), *Learning Machines*, McGraw-Hill, New York.

Olson, E. C., and Christofferson, R. E. (1979), *Computer Assisted Drug Design*, ACS Symposium Series 12, American Chemical Society, Washington, D.C., p. 112.

Overall, J. E., and Kleh, C. J. (1972), *Applied Multivariate Analysis*, McGraw-Hill, New York.

Perrin, C. L. (1974), "Testing of computer-assisted methods for classification of pharmacological activity," *Science*, Vol. 183, pp. 551-552.

Press, S. J. (1972), *Applied Multivariable Analysis*, Holt, Rinehart and Winston, New York.

Purcell, W. P., Bass, G. E., and Clayton, J. M. (1973), *Strategy of Drug Design: A Guide to Biological Activity*, Wiley, New York.

Randic, M. (1978), "Graph-theoretical analysis of structure-property and structure-activity correlations," *Int. J. Quantum Chem., Quantum Biol. Symp.*, Vol. 5, pp. 245-255.

Redl, G., Cramer, R. D., and Berkoff, C. E. (1974), "Quantitative Drug Design," *Chem. Soc. Rev.*, Vol. 28, pp. 273-292.

Richards, W. G., and Black, M. E. (1975), "Quantum chemistry in drug research," *Prog. Med. Chem.*, Vol. 11.

Robbins, H., and Monro, S. (1951), "A stochastic approximation method," *Ann. Math. Statist.*, Vol. 22, No. 1, pp. 400-407.

Robinson, A. B., and Pauling, L. (1974), *Clin. Chem.*, Vol. 20, p. 961.

Rohlf, F. J. (1977), "A probabilistic minimum spanning tree algorithm," *IBM Research Report C 6502*.

Sacco, W., Ashman, W. P., Broome, P. H., and King, J. (1978), *Introduction to Pattern Recognition Applications*, DHEW Publ. FDA 78-1046, Struct. Correl. Carcinog. Mutagen.

Sammon, J. W. (1969), "A nonlinear mapping for data structure analysis," *IEEE Trans. Comput.*, Vol. C-18, pp. 401-409.

Schoenfeld, P. S., and DeVoe, J. R. (1976), "Statistical and mathematical methods in chemistry," *Anal. Chem.*, Vol. 48, p. 403R.

Seber, G. A. F. (1977), *Linear Regression Analysis*, Wiley, New York.

Sebestyen, G. S. (1962), *Decision-Making Processes in Pattern Recognition*, Macmillan, New York.

Singer, J. A., and Purcell, W. P. (1967), "Relationships among current quantitative structure-activity models," *J. Med. Chem.*, Vol. 10, pp. 1000-1002.

Smith, E. G., and Baker, P. A. (1976), *The Wiswesser Line-Formula Chemical Notation (WLN)*, 3rd ed., CIMI, Cherry Hill, N.J.

Sneath, P. H. A. (1966), "Relations between chemical structure and biological activity in peptides," *J. Theor. Biol.*, Vol. 12, pp. 157-195.

Snedecor, G. W. (1956), *Statistical Methods*, 5th ed., Iowa State College Press, Ames, Iowa.

Soltzberg, L. J., and Wilkins, C. L. (1976), "Computer recognition of activity class from molecular transforms," *J. Am. Chem. Soc.*, Vol. 98, pp. 4006-4011.

Soltzberg, L. J., et al. (1976), "Evaluation and comparison of pattern classifiers for chemical applications," *J. Am. Chem. Soc.*, Vol. 98, pp. 7139-7144.

Spendley, W., Hext, G. R., and Himsworth, F. R. (1962), "Sequential application of simplex designs in optimization and evolutionary operation," *Technometrics*, Vol. 4, No. 4, pp. 441-461.

Srivastaver, M. S., and Khatri, C. G. (1979), *An Introduction to Multivariable Statistics*, North-Holland, Amsterdam.

Stuper, A. J. (1977), "Studies of the relationships between drug structure and biological activity using pattern recognition techniques," Ph.D. thesis, The Pennsylvania State University.

Stuper, A. J., and Jurs, P. C. (1975), "Classification of psychotropic drugs as sedatives or tranquilizers using pattern recognition techniques," *J. Am. Chem. Soc.*, Vol. 97, pp. 182-187.

Stuper, A. J., Brugger, W. E., and Jurs, P. C. (1979), *Computer Assisted Studies of Chemical Structure and Biological Function*, Wiley, New York.

Sybrandt, L. B., and Perone, S. P. (1971), "Computerized learning machine applied to quantitative analysis of mixtures by stationary electrode polarography," *Anal. Chem.*, Vol. 43, pp. 382-388.

Tal'Roze, V. L., Reznikov, V. V., and Tantsyver, G. D. (1964), "Minimum information sufficient for identification of individual organic substances by the coincidence of their mass spectral lines," *Dokl. Akad. Nauk SSSR*, Vol. 159, p. 182.

Tatsuoka, M. M. (1971), *Multivariate Analysis: Techniques for Educational and Psychological Research*, Wiley, New York.

Thompson, W. O., and Cady, F. B., eds. (1973), *Proceedings of the University of Kentucky Conference on Regression with a Large number of Prediction Variables*, Department of Statistics, University of Kentucky, Lexington, Kty.

Timm, N. H. (1975), *Multivariable Analysis with Applications in Education and Psychology*, Brooks/Cole, Monterey, Calif.

Ting, K. H., Lee, R. C., Milne, W. A., Shapiro, M., and Guarino, A. M. (1973), "Applications of artificial intelligence: relationships between mass spectra and pharmacological activity of drugs," *Science*, Vol. 108, p. 417.

Topliss, J. G. (1972), "Utilization of operation schemes for analog synthesis in drug design," *J. Med. Chem.*, Vol. 15, p. 1006.

Topliss, J. G., and Edwards, R. P. (1979), "Chance factors in studies of quantitative structure-activity relationships," *J. Med. Chem.*, Vol. 22, No. 10, pp. 1238-1244.

Tou, J. T., and Gonzalez, R. C. (1974), *Pattern Recognition Principles*, 2nd ed., Addison-Wesley.

Usdin, E., and Efron, D. H. (1972), *Psychotropic Drugs and Related Compounds*, DHEW Publ. (HSM) 72-9074.

Wangen, L. E., and Isenhour, T. L. (1970), "Semiquantitative analysis of mixed gamma-ray spectra by computerized learning machines," *Anal. Chem.*, Vol. 42, pp. 737-743.

Watanabe, S. (1969), *Methodologies of Pattern Recognition*, Academic Press, New York.

White, R. F., and Lewinson, T. M. (1977), "Probabilistic clustering for attributes of mixed type with biopharmaceutical application," *J. Am. Statist. Assoc.*, Vol. 72, p. 271.

Whitney, V. K. M. (1972), "Algorithm 422, minimal spanning tree," *Commun. ACM*, Vol. 15, pp. 273-274.

Wilkins, C. L., and Isenhour, T. L. (1975), "Multiple discriminant function analysis of $C^{13}$-NMR spectra: functional group identification by pattern recognition," *Anal. Chem.*, Vol. 47, pp. 1849-1851.

Wilkins, C. L., Williams, R. C., Brunner, T. R., and McCrombie, P. J. (1974), "Heuristic pattern recognition analysis of carbon-13 nuclear magnetic resonance spectra," *J. Am. Chem. Soc.*, Vol. 96, pp. 4182-4185.

Wooldridge, K. R. H. (1980), "Computers and the medicinal chemist," *Chem. Ind.*, Vol. 21, pp. 478-482, (June).

Wootton, R., Cranfield, R., Sheppey, G. C., and Goodford, P. J. (1974), "Physicochemical-activity relationships in practice: 2. Rational selection of Benzenoid substituents," *J. Med. Chem.*, Vol. 18, No. 6, pp. 607-612.

Zahn, C. T. (1971), "Graph-theoretical methods for detecting and describing Gestalt clusters," *IEEE Trans. Comput.*, Vol. C-20, No. 1, pp. 68-86.

Zander, G. S., and Jurs, P. C. (1975), "Generation of mass spectra using pattern recognition techniques," *Anal. Chem.*, Vol. 47, pp. 1562-1572.

Zander, G. S., Stuper, A. J., and Jurs, P. C. (1975), "Nonparametric feature selection in pattern recognition applied to chemical problems," *Anal. Chem.*, Vol. 47, No. 7, pp. 1085-1093.

# 2

## SCREENING PROCEDURES FOR DISCOVERING ACTIVE COMPOUNDS

### 2.1. INTRODUCTION

QSAR studies and the biological screening of compounds are complementary activities. While the former suggests which compounds to investigate, the latter attempts to ascertain the actual activity of compounds in a biological system. Accordingly, the correct design of biological screening procedures is a vital step in the discovery and development of successful drugs.

How a good screen should be constructed depends very much on the type of information that is to be obtained. A screen that seeks to discover the interesting compounds among a large collection of untested compounds is different from a screen that seeks to determine the activity of a specific compound to a specified degree of accuracy. These, in turn, are different from screens that seek to discover the most active compound in a small collection of active compounds.

Statistical methodologies pertaining to each of the foregoing types of screen have been discussed in the literature. An extensive but dated bibliography on screening and selection has been compiled by Federer (1963). Standard techniques for determining the activity of a single compound to a specified degree of accuracy is found in many texts (e.g., Bliss, 1967). A related topic, the determination of sequential designs for destructive life testing and animal serial sacrifice experiments has been discussed by Bergman and Turnbull (1983). In addition, procedures for selecting the best among a collection of candidates, as pioneered by Bechhofer, have been recently reviewed by Dudewicz (1980).

Our interest in biological screening procedures is restricted primarily to those techniques that are appropriate for the exploratory phase of pharmaceutical research. The literature in this area is not extensive, the major articles having been written between the years 1958 and 1964. From our interviews with laboratory personnel we have found that little or no use is presently made of these or other statistical techniques during this important initial phase of pharmaceutical research. This is unfortunate, for a considerable improvement in efficiency could be obtained with a small amount of attention paid to the statistical aspects of designing preliminary screening procedures.

In the following sections we formulate the problem in detail, review the methods that have been suggested, and contribute one of our own. In the remainder of this section we give a brief nontechnical review of the methods discussed, and the advantages and disadvantages of each.

Together, all the proposed methods share the following common understanding of the problem. The true activity of a compound in a given biological system is observed only with error. Consequently, two types of errors are apt to occur in judging a compound. First, a highly potent compound may be wrongly rejected because the random disturbances cause the test score to show an uncharacteristically low activity. Second, an inactive compound may be accepted for further studies for precisely the opposite reason. Both of these results are undesirable.

All of the proposed methods also utilize the technique of replication to achieve a solution to this problem. In consequence, the designs of the proposed screen are also of the same general form. They consist of a sequence of stages each characterized by a fixed number of replications, and a collection of score threshold values describing whether the observed compound is to be rejected, passed on for secondary studies, or passed on to the next stage of the screen for additional replications.

The primary differences between the proposed methods are in how they analyze the problem, and hence ultimately in the details of the resulting "best" screen. These details include the optimal number of replications and the optimal threshold values given at each stage.

The simplest of the proposed methods is by Armitage and Schneiderman (1958). As they point out, for any screen there is associated an operating characteristic (OC) curve $\rho(\theta)$ which is defined as the probability that a compound of activity $\theta$ will be passed by that screen. They suggest that contending screens may be compared by comparing their OC curves. Exactly how this comparison is to be made is left to the discretion of the researcher. An example consisting of two different screens is used to illustrate their method.

Their suggestion is not without merit. Usually, a researcher will have some intuitive notion about which true levels of activity he or she would like a screen to pass, and which levels he or she would like to have rejected. The usefulness of the OC curve resides in the fact that it provides exactly this type of information. Consequently, the researcher proceeds by specifying a number of contending designs, computes their corresponding OC curves, and selects the one that gives the required degree of discrimination. The input is minimal. The researcher need only know the degree of error of his or her observations. This may be obtained by making a number of initial replications. The primary assumption is that the experiments are independent, something one would hope would be true. However, the computational effort may in practice be large. Often it is difficult to find analytic expressions for the OC curve, and it may be necessary to compute it by simulation. However, virtually all the methods discussed rely on the OC curve, and therefore this difficulty does not put the Armitage-Schneiderman method at a disadvantage compared to the other methods.

Another function, not mentioned by Armitage and Schneiderman but also useful in this context, is the average sample number function $\eta(\theta)$. It describes the average number of observations required by a given screen to pass judgment on a compound of activity $\theta$. Hence it may be used as a measure of the relative efficiency of the screen at various $\theta$ values. In general, the most efficient screens for a given level of discrimination are fully sequential screens in which there are infinitely many stages. This efficiency decreases as the number of stages decreases. Hence, the single-stage screen is the least efficient. However, the administrative (as distinct from the sampling) cost of a screen increases as the number of stages increases. The latter cost will often outweigh the increased efficiency, and many authors argue (e.g., Colton, 1963, and King, 1963) that three stages are sufficient for practical purposes. In general, the average sample number function is no more difficult to compute than the OC curve. Hence these should be used together to judge contending screen designs.

The remaining papers all go one step further and propose an optimality criterion for selecting among competing designs.

Virtually in consequence, it is assumed in all of these papers that there is a known prior distribution $g(\theta)$ describing the activity distribution among the unknown compounds. This may seem an objectionable feature, but need not present undue difficulty. First, it is rarely that the researcher knows nothing about the activity distribution in his supply. Second, the optimal designs are generally not sensitive to the exact distribution given. Thus even a large error in the specification of $g(\theta)$ may not alter the optimal design. However, it should at the same time be noted that the following assumption accompanies a known prior: This prior applies to all the compounds being submitted to the screen. Hence it is implicitly assumed that the researcher is engaged in a search, choosing randomly among compounds, rather than trying to optimize a lead. Except in extreme cases, violation of this assumption is unlikely to disturb seriously the optimality of a given procedure.

In addition, many of the remaining papers illustrate their criterion by applying it to the following elementary model. First, there are only two types of compounds, actives with activity level $\theta_A$ and inactives with activity level $\theta_B$. This implies that the prior distribution is characterized by the single parameter p describing the proportion of actives in the supply. Second, the error distribution associated with a single replication is normally distributed with constant variance $\sigma^2$.

This specific model has the advantage that it is analytically tractable. In particular, the OC curve and the average sample number function may be expressed as functions of the multivariate normal distribution. King (1963) claims that the two-point distribution provides a reasonably good approximation to the exponential prior that is apt to hold in a laboratory setting. It should be noted that even if this model is not found to be adequate, one can usually generalize the methods proposed to more appropriate models.

In his paper, Colton (1963) proposes the following optimality criterion. Specify limits $\alpha_0$ and $\beta_0$ on the tolerated errors of misclassifying a true negative and true positive, respectively; select the number of stages to be in the design; and then solve for the design that minimizes the average number of observations required while meeting these specifications. Colton derives appropriate formulas to be used in obtaining the number of replications and cutoff values for the previously mentioned standard model when the design consists of one, two, and infinitely many stages. In addition, tables are provided for a collection of different $\alpha_0$ and $\beta_0$ values.

The primary weakness of Colton's optimality criterion is the need to specify $\alpha_0$ and $\beta_0$. For the range of values tabulated, the resulting design is apt to swamp the follow-up studies with a large number of false negatives. However, if reasonable $\alpha_0$ and $\beta_0$ are known a priori, the Colton method may be applied.

Davies (1958) proposed two different criteria based on maximizing per unit effort the number of actives delivered to the follow-up screen given certain constraints on resources. Specifically, he suggested: (1) "reduce the number of compounds to a given fraction in such a way as to maximize the number of active compounds for a given amount of effort," and (2) "specify the proportion of positives to false positives emerging from the test, or the degree of concentration of positives, required and maximize the number of active compounds for a given amount of effort." Davies gives the general form of the constraint equation and the function to be maximized. No analytic formulas are provided that will define the optimal plan. Rather, it is suggested that the optimal design be found by trial and error, exploring the response surface of the function to be maximized with different screen parameters. The Davies approach is applicable when the constraints are of the type specified. Of these two constraints the first seems the more natural.

King (1963) proposes a slightly different criterion. He assumes that there are constraints on both the primary and secondary screens. The first is in terms of the number of animals available per time unit, and the second in terms of the number of compounds that the secondary screen can accommodate per time unit. Given these constrains, he suggests that the proportion of unknowns possessing interesting activity among those accepted be maximized. To assist the researcher who is using the standard model, some analytic formulas are given which define the constraint equation and the function to be maximized. This function must then be maximized by a search algorithm.

In many respects King's objective is not unreasonable. However, as Davies (1963) points out, when the number of compounds to be tested per time unit can be varied by altering the design of the screen, the cost of compounds per time unit will also vary and these costs should be taken into consideration. This fact is not incorporated into King's model.

Davies (1963) proposes a model based on balancing the reward of discovering a marketable drug against the cost of testing, given a restriction on the number of animals that can be consumed per year in both the primary and secondary screens. Beyond the error distribution and the constants defining the constraint, the following inputs are required: V (value of a marketable drug), $C_1$ (cost of a replication), $C_2$ (cost of a compound), $P_1$ (relative frequency of actives), $P_2$ (probability that an active will become a marketable drug), A (competitive effort), h (overlap in effort between the firm and its competitors), and $\gamma$ (annual discount rate on the value of money). Using these inputs the expected cash flow is determined for any screen as a function of its operating characteristic curve and average sampling number function. The objective is then to maximize this cash flow by choosing among those designs that satisfy the constraints. No analytic formulas are given which define the solution, and it is clear that the optimum is to be found by a trial-and-error search routine. Davies does, however, illustrate the results with a simple example. It reveals that the optimum is relatively robust with respect to the parameters that are the most difficult to estimate: V, the prior probability distribution, and the product $P_1 P_2$. Consequently, this model may be useful for the type of constrained optimization situation envisioned.

The idea of balancing rewards and costs has received considerable attention in the closely allied field of acceptance sampling. In a model by Hald (1960) it is assumed that for any level of activity $\theta$, one may identify two different types of losses: the loss $A(\theta)$ experienced when a compound of activity level $\theta$ is accepted, and the loss $R(\theta)$ experienced when such a compound is rejected. The former may represent the cost of additional testing, and if the compound is marketable, the negative of its commercial value as a drug. The latter may represent the forgone commercial value of the compound. In addition, one defines for any contending screen the cost $C(\theta)$ of passing judgment on a compound of activity $\theta$. Given a known prior distribution $g(\theta)$, one may then compute the expected loss of the screen. Hald suggests that the objective should be one of finding that screen which achieves the smallest expected loss. Many specific examples of screens optimal in this sense can be found in the acceptance sampling literature (see Wetherill, 1975). However, none are specifically tailored to a pharmaceutical setting.

Dunnett (1961) applied a variant of the foregoing approach to the standard two-point model with normal error distribution. Rather than the *expected loss of a screen* per *tested compound*, he considers the *expected regret* of a screen per *accepted active compound*. The notion of regret is similar to that of loss, with the exception that the losses are standardized such that the best decision at each (known) $\theta$ value will have a loss of

zero.  Dunnett obtains detailed formulas useful for obtaining
the parameters defining the optimal one-, two-, and three-stage
screens for the standard two-point model.  In addition, a set of
tables are provided.

Although quite interesting, each of the preceding two
models may be criticized.  Both require that the loss $R(\theta)$ due
to rejecting a compound of activity $\theta$ be specified.  This can be
quite difficult.  In commenting on this task, Colton asks rhe-
torically: "How can one assess the cost of missing a possible
anti-cancer drug relative to the cost of experimentation?"

Finally, one of the present authors (Bergman, 1981) has
developed a model for the situation when only a small proportion of all compounds tested can be accepted for further investigation.  It is assumed that the user can specify the rewards
from accepting a compound with true activity $\theta$.  This consists
of the expected market value of such a compound less testing
and development costs, and hence is simply the negative of
$A(\theta)$.  In addition, it is assumed that the costs $C(\theta)$ of employing a screen can be specified.  These are the sampling and setup costs per compound.  Finally, a prior distribution $g(\theta)$ is
assumed.  The objective is now defined to be one of discovering
that screen which will maximize the expected reward of each
compound accepted.  This is similar, but not equivalent to the
objective proposed by Dunnett.  Implicit in this new formulation
is the notion that the loss due to rejecting a compound is the
cost that must be expended in finding a compound of equal activity—an evaluation consistent with the idea that with sufficient effort one can find analog drugs to ones already successfully marketed by a competitor.  This loss is computed internally when solving the functional equation defining the optimal
solution, and thus need never be specified by the researcher.

To aid the user in finding the optimal design, a general
algorithm is given.  In practice, it will be necessary for the researcher to adapt it to the specific model characterizing the error distribution of his or her biological experiments.

Some examples are given which illustrate the types of results one is apt to obtain.  From these results one may observe
that the most difficult parameters to specify, the prior distribution g and the value $V(\theta)$ of a marketable drug having activity
$\theta$, do not appreciably influence the optimal design, and hence
need not be specified with any great precision.  Consequently,
this last model may be useful in determining the optimal screen
when only a small proportion of all compounds can be accepted
for further studies.

## 2.2. FORMULATION OF THE PROBLEM

It is well known that many thousands of compounds are tested for activity for each compound that becomes a marketable drug. Arnow (1970) estimates that 175,000 substances are subjected to biological evaluation each year, of which about 20 become drugs. The principal problem of initial screening is one of reducing this number to a smaller set of interesting compounds which should undergo more intensive testing.

If the initial screen test methodology specified the activity of a compound in a given biological system without error, the problem of selecting which should be tested more intensively would be relatively simple: Select all compounds whose test activities are above a predetermined level. Exactly what level this should be may be a difficult question. It is one we shall return to later in the chapter, but for the present we may assume that it is known. If we now assume, as is the case in practice, that the measurements are confounded with random disturbances, two types of error are apt to occur in our selection procedure. First, we wrongly reject some highly potent compounds because the random disturbances cause the test score to show an uncharacteristically low activity for these compounds. Second, we wrongly accept some inactive compounds for precisely the reverse reason. Both of these results are highly undesirable. In the former we forgo discovering a compound that could have been instrumental in leading to a cure of a disease. In the latter, additional resources are expended on discovering that this was not an interesting compound after all.

Knowledge of the magnitude of the random disturbances may be obtained by repeated testing of the observed activity of some compounds. These disturbances may be correlated with the individual compounds or with the degree of activity possessed by the compounds. As a first approximation it is convenient to assume that the errors are independent of the compound administered. Thus our observed activity $Y_i$ on one administration of the test procedure on compound i may be decomposed as follows:

$$Y_i = \theta_i + \varepsilon$$

where $\theta_i$ is the true activity of compound i with respect to the biological system on which the compound is being tested, and $\varepsilon$ is an error term with mean 0 and some variance $\sigma^2$.

By replicating the test procedure sufficiently often, the error with which we estimate $\theta_i$ may be reduced to any desired

level. In the literature one finds a number of different replication designs.

The simplest of these designs is the single-stage screen in
which all compounds are given n replications and only those
compounds whose mean score is, say, greater than some
threshold value k are passed. Alternatively, if one is measuring, say, reduction in tumor size, all compounds whose mean
score is less than some value k are passed. In both cases the
question of interest is what values should be given to n and k.

More complex are the multiple-stage screens in which compounds must pass through a succession of basically similar
screens. Typical of these designs is a two-stage screen in
which the first stage consists of $n_1$ replications and the second
stage of $n_2$ replications. All compounds enter the first screen,
and only those that achieve a threshold value of $k_1$ are passed
onto the second screen; there, in the second screen, the passed
compounds undergo $n_2$ additional replications, and only those
that achieve a combined result higher than a threshold $k_2$ are
finally passed. The question to be answered in this context is
what values should be given to $n_1$, $n_2$, $k_1$, and $k_2$. A variation of this design may be obtained by permitting compounds to
forgo the second stage if their score on the first screen is sufficiently high (i.e., higher than a threshold $h_1$). In this case
the additional question arises of what value should be given
to $h_1$.

The advantage of the two-stage screen over the one-stage
screen is that the same degree of discrimination between interesting and uninteresting compounds can be achieved with a
smaller total number of observations. The reason is simple.
Those compounds that show no promise whatsoever are rejected
with relatively few observations at the first screen. Only the
relatively promising ones are given a full complement of replications. A disadvantage, however, is that the procedure is more
complicated, and hence requires more administrative effort.

The two-stage procedure easily generalizes to three or more
stages. In the limit one obtains the fully sequential testing
procedure, where a compound is judged after each replication
as to whether to reject, continue for more replications, or pass
immediately. The basic design configuration in this case is a
sequence of boundaries $(k_i, h_i)$ (i = 1, 2, ...) such that the
compound is rejected if on the ith screen the combined score is
less than $k_i$, is accepted if this score is greater than $h_i$, and
is continued if the score is between these values. By a suitable transformation one sometimes finds that there are two constants k and h such that $k_i = k$ and $h_i = h$ for all i.

Clearly, the fully sequential screen will be the most efficient in terms of the average total number of tests required per compound for a given level of discrimination. However, it is the most expensive to administer. Both Davies (1963) and Armitage and Schneiderman (1958) point out that three stages are usually sufficient for any effective test.

The papers that are reviewed all address screens of this type. They differ in the circumstances in which the screen is to be employed and in the criteria used for selecting the number of replications and the cutoff values. These papers do not, however, deal with a number of other questions associated with screens of this type. These include crossover designs, covariance analysis, and other statistical techniques to eliminate much of the biological variation between tests. We shall assume that such methods are employed and that the resultant reduced variability per test is the $\sigma^2$ mentioned previously. For further literature on this aspect of design, see Bliss (1967).

## 2.3. THE OC CURVE AND THE AVERAGE SAMPLE NUMBER

In two early papers on the subject of pharmaceutical screens, Davies (1958) and Armitage and Schneiderman (1958) both observed that compound screening has similarities with acceptance sampling. In the latter, one samples and tests batches of a product to see which are of acceptable quality and which should be rejected. Each item selected from a batch yields an additional observation on the underlying quality of the batch, and corresponds to one of a sequence of replicate tests on a compound.

A principal tool in acceptance sampling for designing and evaluating screening procedures is the operating characteristic (OC) curve $\rho(\theta)$, which gives the probability of accepting a batch (compound) of any given quality (activity) $\theta$ under some specified sampling plan. Borrowing from the methodology of acceptance sampling, Armitage and Schneiderman consequently suggest the following approach to compound screening: For any given screening procedure, compute the OC curve of the procedure and decide if it gives sufficiently high certainty that interesting compounds are passed and uninteresting ones are not.

Armitage-Schneiderman give the following example of a three-stage screen. A group of animals treated with a compound is tested against a control group for reduction of a tumor.  At each successive stage of the procedure the quantity log $T_i/C_i$ ($i = 1,2,3$) is computed, where $T_i$ is the mean tumor size of the treated group at the ith stage, and $C_i$ is defined similarly for the control.  Two schemes are considered, which are summarized below.

|  | Scheme | |
| --- | --- | --- |
|  | A | B |
| After stage 1, reject if<br>$\log T_1/C_1 >$<br>Otherwise, continue to stage 2 | log 0.54 | log 0.60 |
| After stage 2, reject if<br>$\log T_1/C_1 + \log T_2/C_2 >$<br>Otherwise, continue to stage 3 | log 0.20 | log 0.30 |
| After stage 3, reject if<br>$\log (T_1/C_1) + \log (T_2/C_2)$<br>$+ \log (T_3/C_3) >$<br>Otherwise, accept | log 0.08 | log 0.15 |

By repeated sampling it was determined that the replication error for log T/C at each stage was approximately normally distributed with standard deviation log 0.13.  The corresponding operating characteristic curves are displayed in Figure 18. Both of these schemes guarantee that a compound with true activity log 0.2 (i.e., an inhibition of 80%) will almost certainly be accepted, while a compound with activity log 0.8 (i.e., an inhibition of only 20%) will not be accepted.  In between these values, scheme B will accept more compounds, decreasing the number of false negatives but increasing the number of false positives.  The appropriate scheme will depend on which values one wants to screen for or against.

One of the useful features of the operating characteristic curve is that it may be used to compare any pair of sampling plans with respect to the probability of passing a compound of any given activity.  However, as any level of certainty can be

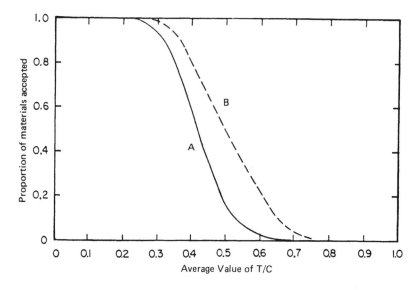

**Figure 18** Three-stage scheme, A, contrasted with a less severe 3-stage scheme, B. T/C is the ratio of mean tumor weights in the test and control groups.

achieved by taking sufficiently many replications, this in itself is probably insufficient for drawing conclusions about the value of contending procedures. Also needed is the average sample number $\eta(\theta)$, defined as the expected number of observations needed under the procedure to either reject or accept a compound of activity $\theta$.

In principle, both the OC curve $\rho(\theta)$ and the average sample number $\eta(\theta)$ may be computed for any procedure. However, in practice, the task may be difficult, and sometimes it is necessary to resort to simulation. For illustrative purposes we shall show how to compute these functions for a two-stage plan, under the assumption that if $\theta$ is the true activity of a compound, the corresponding ith stage observation $X_i$ ($= \log T_i/C_i$) is normally distributed with mean $\theta$ and variance $\sigma_i^2$. As in the Armitage-Schneiderman example, it is assumed that a compound will be rejected if $Y_i = (1/i) \Sigma_{j=1}^{i} X_j$ exceeds some value $k_i$ at stage i (i = 1,2).

Since, by definition, the OC curve at $\theta$ is the probability of accepting a compound of activity $\theta$ under the plan, it follows that for two-stage plans defined as above,

$$\rho(\theta) = P(Y_1 < k_1, Y_2 < k_2 | \theta)$$

Setting

$$Z_1 = \frac{Y_1 - \theta_1}{\sigma_1} \quad \text{and} \quad Z_2 = \frac{2(Y_2 - \theta_2)}{(\sigma_1^2 + \sigma_2^2)^{1/2}}$$

we note that the distribution of $X_1$ and $X_2$ implies that $(Z_1, Z_2)$ have a bivariate normal distribution with mean $(0,0)$ and covariance matrix

$$\begin{bmatrix} 1 & r \\ r & 1 \end{bmatrix}$$

where

$$r = \frac{\sigma_1}{(\sigma_1^2 + \sigma_2^2)^{1/2}}$$

Consequently, if $G(x,y,r)$ denotes the standardized bivariate normal distribution function with correlation coefficient $r$, it then follows that

$$\rho(\theta) = P\left( Z_1 < \frac{k_1 - \theta}{\sigma_1}, \ Z_2 \leqslant \frac{2(k_2 - \theta)}{(\sigma_1^2 + \sigma_2^2)^{1/2}} \ \Big| \ \theta \right)$$

$$= G\left( \frac{k_1 - \theta}{\sigma_1}, \ \frac{2(k_2 - \theta)}{(\sigma_1^2 + \sigma_2^2)^{1/2}}, \ \frac{\sigma_1}{(\sigma_1^2 + \sigma_2^2)^{1/2}} \right)$$

Thus the OC curve may be obtained directly from bivariate normal distribution tables.

Similarly, the average sample number $\eta(\theta)$ is the expected number of observations needed under the procedure to either accept or reject the compound. Consequently, if we assume that the first stage of the plan requires $n_1$ test animals and the second stage requires $n_2$, then

$$\eta(\theta) = n_1 + n_2 P(Y_1 < k_1 | \delta)$$

$$= n_1 + n_2 \Phi\left(\frac{k_1 - \theta}{\sigma_1}\right)$$

where $\Phi(\cdot)$ is the standard univariate normal distribution function.

The example above may be generalized. First, the assumption that $\sigma_1^2$ does not vary with $\theta$ is easily dispensed with by substituting $\sigma_i^2(\theta)$ for $\sigma_i^2$, where $\sigma_i^2(\theta)$ is the variance expressed as a function of $\theta$. Second, one may derive the m-stage OC curve and average sample number by considering the m-dimensional multivariate normal distribution. Third, distributions other than the normal may be assumed. For example, the binomial distribution is obtained if what one observes is the number of positive responses in a group of n animals. A paper by Schultz et al. (1973) gives an explicit computer-adaptable algorithm for calculating $\rho(\theta)$ and $\eta(\theta)$ for any multistage plan when the observations at each stage constitute a binomial distribution.

## 2.4. PROCEDURES FOR LIMITING MIS-CLASSIFICATION ERRORS

Colton (1963) points out that the OC approach to compound screening does not actually solve the problem of selecting among various contending screens. It informs the user about the characteristics of these plans and leaves it to him or her to decide the relative importance of these characteristics, and hence which plan is most suited to his or her needs.

To define a criterion by which to rate sampling plans more formally, Colton draws on the classical notions found in the Neyman-Pearson theory of hypothesis testing. This criterion may be illustrated by the simple example in which there are exactly two kinds of compounds: actives with parameter $\theta_A$ and inactives with parameter $\theta_B$. In this case the OC curve of any given plan will consist of two points $\rho(\theta_A)$ and $\rho(\theta_B)$, or equivalently, the two quantities $\alpha$ and $\beta$, defined by

$$1 - \alpha = \rho(\theta_A)$$

$$\beta = \rho(\theta_B)$$

(1)

which are the respective probabilities of accepting an active compound, and of accepting an inactive compound, under the plan.

As Colton suggests, a plausible objective would be to select a plan which guarantees that $\alpha$ and $\beta$ are suitably low, say less than $\alpha_0$ and $\beta_0$. If there is more than one plan satisfying the constraints

$$\alpha \leqslant \alpha_0$$

$$\beta \leqslant \beta_0$$

(2)

then the one for which $\eta(\theta)$ is smallest should be chosen. However, $\eta(\theta)$ is a function evaluated at two points, $\theta_A$ and $\theta_B$, and consequently no such minimum may exist. To overcome this difficulty, Colton assumes that there exists a known prior probability p (= 1 − q) corresponding to the proportion of actives among all compounds. Consequently, the objective is now to determine that plan which satisfies (1) and minimizes the expected number of observations given p, that is, minimizes

$$N = p\eta(\theta_A) + q\eta(\theta_B)$$

(3)

The theory for this situation is worked out for a number of different screens under the assumption that in a single replication we observe

$$Y_i = \theta_i + \varepsilon$$

where $\theta = \theta_A$ or $\theta_B$ and $\varepsilon$ is an error term which is normally distributed with mean 0 and known variance $\sigma^2$.

For a single screen with n replications and cutoff value k [i.e., accept if $\bar{X} = (1/n)\Sigma_{i=1}^{n} X_i < k$], equations (1) are equivalent to

$$1 - \alpha = \Phi\left(\frac{k - \theta_A}{\sigma/\sqrt{n}}\right)$$

$$\beta = \Phi\left(\frac{k - \theta_B}{\sigma/\sqrt{n}}\right)$$

The minimizing N in (3) is given by the solutions n and k of these equations. If $Z_\gamma$ is the solution of the equation $\gamma = \Phi(Z_\gamma)$, one obtains

$$n = \frac{(Z_{1-\alpha} - Z_\beta)^2}{\delta^2}$$

$$k = \frac{(Z_{1-\alpha}\theta_B - Z_\beta\theta_A)}{(Z_{1-\alpha} - Z_\beta)}$$

where $\delta = (\theta_B - \theta_A)/\sigma$.

Colton also considers the two-stage plan of the form

*Stage 1*: Take $n_1$ observations. Reject if $\bar{X}_1 > k_1$; otherwise, continue.

*Stage 2*: Take $n_1$ observations. Reject if $(n_1\bar{X}_1 + n_2\bar{X}_2)/(n_1 + n_2) > k_2$; otherwise, accept.

In this case

$$\rho(\theta) = G\left(\frac{k_1 - \theta}{\sigma n_1^{-1/2}}, \frac{k_2 - \theta}{\sigma(n_1 + n_2)^{-1/2}}, \left(\frac{n_1}{n_1 + n_2}\right)^{1/2}\right)$$

$$\eta(\theta) = n_1 + n_2\Phi\left(\frac{k_1 - \theta}{\sigma/\sqrt{n_1}}\right)$$

Thus (3) is minimized subject to (2) [or equivalently, (1)]. For this problem Colton provides tables that describe the optimal $n_1$, $n_2$, $k_1$, and $k_2$ values for some selected $\alpha_0$, $\beta_0$, and p values. For these values Colton finds that imposing the restriction $n_1 = n_2$ does not increase the minimum of (3) by more than 3%.

Finally, the optimal parameters of the fully sequential plan are also obtained. This plan is of the following form. Set $\bar{X}_m = (\sum_{i=1}^m X_i)/m$, where $X_i$ is the ith observation, and use the screening rules

$$\bar{X}_m \leqslant k_m \Rightarrow \text{accept}$$

$$\bar{X}_m > h_m \Rightarrow \text{reject}$$

$$k_m < \bar{X}_m \leqslant h_m \Rightarrow \text{continue}$$

Then Wald's sequential probability ratio test uniquely determines the boundaries to be, approximately,

$$k_m = \frac{\sigma^2/m}{\theta_B - \theta_A} \log \frac{\beta_0}{1 - \alpha_0} + \frac{\theta_B + \theta_A}{2}$$

$$h_m = \frac{\sigma^2/m}{\theta_B - \theta_A} \log \frac{1 - \beta_0}{\alpha_0} + \frac{\theta_B + \theta_A}{2}$$

Furthermore, under these boundaries

$$N = \frac{2}{\delta^2} \left\{ [q\beta_0 - p(1 - \alpha_0)] \log \frac{\beta_0}{1 - \alpha_0} \right.$$

$$\left. + [q(1 - \beta_0) - p\alpha_0] \log \frac{1 - \beta_0}{\alpha_0} \right\}$$

Colton's optimality criterion has been criticized by King (1963) on the grounds that it tends to swamp the follow-up study facilities with an excessive number of false positives for the $\alpha_0$ and $\beta_0$ tabulated. This may be easily seen. If 0.1% of the submitted compounds are true positives, $\delta = 1$, and the misclassification probabilities are both set at 0.05, then for the single-stage screen approximately 5% of all compounds are passed, of which approximately 98% are false positives. The effort expended in additional tests discovering that most of these are in fact false positives could very likely have been more profitably used in testing new compounds. This could be avoided by choosing $\alpha_0$ and $\beta_0$ at a more sensible level. However, it is not intuitively obvious what levels these should be.

## 2.5. PROCEDURES FOR MAXIMIZING BENEFITS SUBJECT TO A CONSTRAINT

A different approach to selecting a "best" screen may be based on the recognition that there is an implicit trade-off between the amount of testing that is performed on an individual compound to learn its activity and the number of compounds that can be tested. Davies (1958), drawing on the ideas contained in a private communication from J. T. Lichfield, puts it this way:

For a given total amount of testing effort the number of compounds we test is inversely proportional to the amount of testing per compound. For a given amount of testing per compound there will be a given risk of missing any given active compound. If we reduce the testing per compound, we will increase the risk of missing a given positive, but since we shall be able to test a larger number of compounds, we may end up with a larger number of positives. There are thus two ways of missing positives, one is by falsely classifying them, and the other is by not being able to test them. These are conflicting tendencies because if we reduce the losses in one, we increase the losses in the other. The problem is to find the best balance.

This suggests that one should define an optimum procedure as one that yields the largest number of positives per unit of effort. Clearly, the solution to this problem is to accept all compounds at no sampling cost. Consequently, some additional restrictions are needed to obtain a reasonably well specified problem. Davies identified the following three alternative objectives as worthy of consideration: (1) "reduce the number of compounds to a given fraction in such a way as to maximize the number of active compounds for a given amount of effort"; (2) "reduce the number of compounds to a given fraction in such a way as to maximize the mean activity for a given amount of effort"; and (3) "specify the proportion of positives to false positives emerging from the test, or the degree of concentration of positives required, and maximize the number of active compounds for a given amount of effort." Of these, Davies considered only objectives 1 and 3, regarding objective 2, previously explored by Finney (1958), as inappropriate for the time constraints of pharmaceutical research.

In formulating objectives 1 and 3 Davies assumes that each compound possesses an inherent activity (parameterized by $\theta$) and that the researcher has established that any compound with inherent activity greater than some value a is to be considered interesting or active. In addition, it is assumed that there exists a known frequency function $g(\theta)$ for the activity of all compounds to be tested, and that there exists a known cost function $C(n)$ which describes the cost of testing on n animals.

For any given plan one may define the associated OC curve $\rho(\theta)$ and average sample number function $\eta(\theta)$ (see Section 2.3). Consequently,

$$R = \int_{-\infty}^{\infty} \rho(\theta) g(\theta) \, d\theta$$

is the proportion of the compounds that will be passed under the plan. The proportion that are active is

$$p = \int_{a}^{\infty} \rho(\theta) g(\theta) \, d\theta$$

and the cost of the plan is

$$C = \int_{-\infty}^{\infty} C(\eta(\theta)) g(\theta) \, d\theta$$

In terms of the notation above, objective (1) may now be formulated as maximizing $P/C$ subject to $R = r$, where $r$ is a specified fraction. Objective (3) corresponds to maximizing $P/C$ subject to $P/R = b$, where $b$ is a specified fraction.

Solving for the maximum requires numerical techniques, and Davies discussed in limited detail the two-stage and fully sequential designs when it is assumed that the prior $g(\theta)$ is two-point and the sampling error is normal with constant variance $\sigma^2$. He finds that the maximum is relatively flat with respect to the parameters characterizing the plan ($n_1$, $n_2$, $k_1$, and $k_2$ for the two stage, and $k$ and $h$ for the fully sequential) and that the solution can be found by trial and error using a computer to search the response surface of the function to be maximized.

King, in two papers (1963), (1964), continues the work of Davies. In both papers he considers the problem of maximizing the proportion of unknowns possessing "interesting activity" among those accepted, subject to existing restrictions on available resources. These restrictions encompass both the primary screen and secondary follow-up capacities. Specifically, it is assumed that the primary screening facilities are limited to a maximum of A groups of animals per time period, and that the follow-up (secondary) testing facilities have a maximum capacity of U unknowns per time period. Given a known prior distribution the objective to be maximized in terms of our previous notation is $P/R$ subject to the constraint

$$\frac{AR}{C} = U$$

where $C(\eta(\theta)) = \eta(\theta)$. Setting $\delta = U/A$, the proportion of secondary unknowns to primary groups of animals, this may be rewritten as

maximize $\dfrac{P}{\delta C}$

subject to the constraint

$$R = \delta C$$

King discusses the general problem of finding the optimal values $n_1, \ldots, n_m, k_1, \ldots, k_m$ for the m-stage design. It is assumed that at each stage i the quantity $\bar{X}_i = (1/n_i) \sum_{j=1}^{n_i} X_{ij}$ is recorded, where $X_{ij}$ is the observed activity of the jth group put on the unknown compound at stage i and that $X_{ij}$ is normally distributed with mean $\theta$ and variance $\sigma^2$. With

$$Z_i = \frac{\sum_{j=1}^{i} n_j \bar{X}_j}{\sum_{j=1}^{i} n_j}$$

the rules of the screen are the standard ones:

Reject at stage i if $Z_i < k_i$; otherwise, continue.
Accept at stage m if the compound reaching stage m is not rejected at stage m.

As King observes, solving for $n_1, \ldots, n_m$ and $k_1, \ldots, k_m$ can be quite a formidable task. He notes, however, that if the prior distribution is exponential, as observed in practice by Davies, and if the proportion of interesting compounds is small, the prior may be approximated by a two-point distribution

$$g(\theta) = \begin{cases} p & \text{if } \theta = \theta_A \\ q & \text{if } \theta = \theta_B \end{cases}$$

where $\theta_B = 0$. The problem is then to maximize

$$\frac{p\rho(\theta_A)}{\delta(p\eta(\theta_A) + q\eta(0))}$$

subject to

$$p\rho(\theta_A) + q\rho(0) = \delta(p\eta(\theta_A) + q\eta(0))$$

where $\rho(\theta)$ and $\eta(\theta)$ are the OC curve and average sample number function, respectively. Yet even this is a difficult problem and King suggests that the OC curve

$$\rho(\theta) = P(Z_1 \geqslant k_1, \ldots, Z_m \geqslant k_m | \theta)$$

may be approximated by

$$\prod_{i=1}^{m} P(Z_i \geqslant k_i | \theta)$$

When the proportion of actives in the supply is small, King finds this approximation to work rather well. Charts are then computed which assist the researcher in determining the optimal values for one-, two-, and three-stage designs.

Dunnett (1972) illustrates another variant of Davies' maximization objective. In the example given, Dunnett assumes that there are 10,000 compounds available for testing, of which only 40 are active as antitumor agents. The method of testing is to implant cancer cells into mice and measure the mean difference in tumor weight between treated and control mice.

The primary constraint is on testing facilities, which permit only 50 mice per week to be tested. These may be allocated between initial first-stage tests requiring three mice per compound, or follow-up tests requiring 30 mice per compound. Thus passing a compound to the second stage implies that 10 new compounds are delayed from immediate testing. Assuming that the standard error of the difference between control and treated means is 0.35 g, and that each of the 40 actives is capable of reducing tumor weights by 0.7 g, Dunnett attempts to maximize the proportion of actives found for a given amount of effort by finding the optimal cutoff decision point for passing compounds from the first screen to the second screen. Hence, in this simple variant, what is being optimized is the balance of resources between the two different screens, given that the design of each is fixed except for the passing rule.

By direct computation, Dunnett finds that for a cutoff of 0.7 for the difference between treated and untreated means, the expected number of positives detected among the 10,000 compounds is 247, of which 20 are true positives. The associated equivalent number of first-stage tests is therefore 247 × 10 + 10,000 = 12,470. Thus the number of true actives found per 1000 tests is (20/12,470) × 1000 = 0.160. For a cutoff of a 0.6 difference, one obtains 459 positives, of which 25 are true positives. This leads to a ratio of 0.171 true actives per 1000 tests. Continuing in this fashion, Dunnett obtains a graph of the yield as given in Figure 19. From it, it is evident that the optimal cutoff is near 0.6, and this defines the optimal screen under the stated criterion.

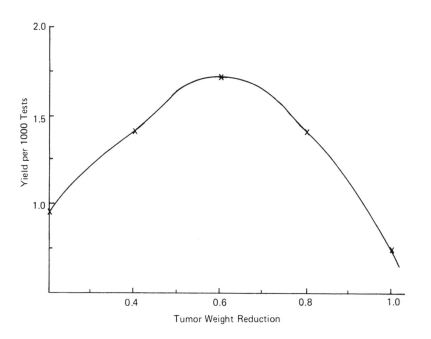

**Figure 19** Yields for screen with different cut-off values. (Reprinted with permission of Holden-Day, Oakland, California (1972).)

## 2.6. PROCEDURES FOR MAXIMIZING NET FINANCIAL BENEFITS

Davies (1963) has pointed out that when the total effort varies with the screen design, some account must be taken of the actual costs involved. One design may yield a greater number of actives per year but only at a greater cost. Davies considers one such model in which the yearly testing effort is fixed in terms of the number of animals used, but not in terms of the number of compounds tested. The problem is one of achieving a balance between the value of discovering a marketable drug and the cost of screening compounds.

For simplicity, Davies considers the case where there are only two types of compounds: actives, with parameter $\theta_A$, and inactives, with parameter $\theta_B$. The relative frequency of the former is $g(\theta_A) = P_1$, and the probability that an active compound will lead to a marketable drug is $P_2$. Hence $p = P_1 P_2$ is the probability that any given compound will lead to a marketable drug. The undiscounted value of such a drug is denoted V, and the associated annual discount parameter is $\gamma$.

The primary restriction is on the number of animals S that may be tested per year. This total consists both of those consumed in the primary screen and those needed in follow-up studies. It is assumed that each compound passing the primary screen will require b additional animals during the course of the secondary screen. In contrast, the number of animals used during the primary screen depends on this screen's design. Davies expresses the screen's design characteristics in terms of the OC curve $\rho(\theta)$ and average sample number function $\eta(\theta)$. For a two-point prior, this gives the parameters $\rho(\theta_A)$, $\rho(\theta_B)$, $\eta(\theta_A)$ and $\eta(\theta_B)$, or, equivalently, $\alpha$, $\beta$, $\eta_A$, and $\eta_B$, where

$$\alpha = \rho(\theta_B)$$

$$\beta = 1 - \rho(\theta_A)$$

$$\eta_A = \eta(\theta_A)$$

$$\eta_B = \eta(\theta_B)$$

Thus, if m is the number of compounds evaluated per year, the constraint on S becomes

$$S \geq m \{ \eta_A P_1 + \eta_B (1 - P_1) + [(1 - \beta)P_1 + \alpha(1 - P_1)]b \}$$

We may assume that for given constants S, b, and $P_1$, and given design parameters $\alpha$, $\beta$, $\eta_A$, and $\eta_B$, the constant m is chosen so that equality is approximately obtained in the inequality above.

The quantity to be maximized is the expected cash flow generated by a screen. Among other things, it is a function of V, $\gamma$, and the expected effective annual effort a. The latter Davies defines as $(1 - \beta)m/f$, where "f is the overlap in research between the firm employing the screen and its competitors." Presumably, $1/f$ may be interpreted as the probability that if the firm finds a marketable drug, it will manage to patent it before its competitors. With this definition for a, the probability of finding a marketable drug in time interval (t, t + dt) is ap dt. Hence the discounted gross expected cash flow is $\int_0^\infty Vape^{-\gamma t} \, dt$. To obtain the net cash flow, the expected cost of effort $\int_0^\infty (C_1 + mC_2)e^{-\gamma t} \, dt$ must be subtracted, where $C_1$ is the annual cost of testing effort and $C_2$ is the cost of each compound. This yields the expression

$$\frac{m(1 - \beta)pV/f - (C_1 + mC_2)}{\gamma}$$

Davies also takes into consideration the lost time from changing from one project to another due to preemption. This is a function of the competitive effort, which is denoted A. The expected time on the project is therefore $E = [(a + A)p]^{-1}$. Letting T represent the average time lost, the proportion of time lost from change over to another project is $T/(T + E)$. Davies finally defines the expected value of a screen to be

$$\left( \frac{E}{T + E} \right) \frac{paV - (C_1 + mC_2)}{\gamma} = \frac{m(1 - \beta)pV/f - (C_1 + mC_2)}{\{1 + T[m(1 - \beta)/f + A]p\}\gamma}$$

The best screen is therefore the screen that maximizes the expression above in $\beta$, and hence also in m, subject to the constraint. Noting that the denominator changes very little for different values of $\beta$ (and m), it is sufficient to maximize the numerator alone subject to the constraint.

The maximization is accomplished through trial and error, exploring the value of various alternative screens. For a single-stage screen $\eta_A$ and $\eta_B$ will be determined uniquely once $\alpha$

and $\beta$ are specified. For a multistage screen the situation is more complex. We may note that the function to be maximized is an increasing function of m. Furthermore, in the constraint equation we see that m may be increased by decreasing $n_A P_1 + n_B(1 - P_1)$. Hence it is clear that for any given $\alpha$ and $\beta$ we need only consider screens that minimize $n_A P_1 + n_B(1 - P_1)$ (i.e., achieve the smallest average amount of sampling while maintaining the specified misclassification errors $\alpha$ and $\beta$). This implies that the alternative screens we may consider are essentially parameterized by $\alpha$ and $\beta$ alone. Achieving this reduced parameterization is clearly not a trivial matter, as the paper by Colton discussed earlier amply demonstrates.

Davies applies his model to the one-stage screen. From the resulting numerical results he observes that the solution is relatively insensitive to the values of the parameters that are most difficult to specify with any accuracy: V, p, and f.

An alternative to the Davies approach is to assign costs or rewards to accepting or rejecting compounds as functions of their activity. This approach has gained a wide following in acceptance sampling. Of interest in this respect is the work of Hald (1960). His model suggests the following formulation for pharmaceutical screens. Let $A(\theta)$ and $R(\theta)$ denote, respectively, the costs of deciding to accept, or reject, a compound of activity $\theta$. For example, $A(\theta)$ may represent the expected loss in additional testing costs, while $R(\theta)$ may be the forgone expected market value of a compound of activity $\theta$. Schematically, these are depicted in Figure 20. We see, in particular, that knowledge of these two loss functions permits one to identify a break-even level below which it is more profitable to reject the compound than to submit it to additional studies. Thus, if the activity $\theta$ of a compound were known, it is clear that with respect to this one compound it would be optimal to accept it for further studies if and only if $\theta > \theta'$. What is also of interest here is that we have for the first time endogenously defined what is meant by an "interesting" compound. It is one with a level of activity at which it is at least as profitable to accept the compound for further studies as to reject it.

In practice, the value $\theta$ will not be known through a single replication, and therefore the ideal policy outlined above is not within the realm of possibility. However, by using replications and multiple stages we can approach this ideal policy to any desired degree. But this is accomplished only at additional costs. The objective must therefore be one of balancing the benefits obtained from replications against their costs.

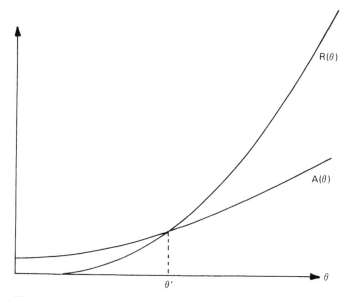

**Figure 20** Costs of acceptance and rejection. (Reprinted with permission of the American Statistical Association.)

For any given screen the net benefits may be computed exactly. If $\rho(\theta)$ and $n(\theta)$ are, respectively, the OC curve and average sample number for the screen, and $C(n(\theta))$ is the cost of applying it, the associated expected benefit (or total cost) as a function of $\theta$ is

$$A(\theta)\rho(\theta) + R(\theta)[1 - \rho(\theta)] + C(n(\theta))$$

With a given known prior distribution $g(\theta)$ for $\theta$, the expected loss is

$$\int_{-\infty}^{\infty} \{A(\theta)\rho(\theta) + R(\theta)[1 - \rho(\theta)] + C(n(\theta))\}g(\theta)\, d\theta$$

Following Hald, a possible objective would be to find the screen that minimizes the expression above. For given $A(\theta)$ and $R(\theta)$, this is equivalent to minimizing the expected loss per compound submitted to the screen, on the assumption that there are no constraints on the number of compounds that can be submitted for additional studies, and that for any n the expected loss of rejecting n compounds of activity $\theta$ is equal to n times the expected loss of rejecting one compound of activity $\theta$. In principle, this last assumption will not hold since only one compound will ultimately be marketed for a given disease. However, if very few compounds are active, the model may work well in practice. Wetherill (1975) has compiled an extensive bibliography on the literature related to such procedures as applied to acceptance sampling. However, to our knowledge, no details have been worked out for specific models suitable for pharmaceutical research.

A variant of the foregoing approach has been put forward by Dunnett (1961). He considers the special case where there are only two types of compounds, actives with activity $\theta_A$ and inactives with activity $\theta_B$. Setting $p = g(\theta_A)$ [and $q = 1 - p = g(\theta_B)$], $C_A = C(\theta_A)$, $C_B = C(\theta_B)$, $\rho_A = \rho(\theta_A)$, $\rho_B = \rho(\theta_B)$, $A = A(\theta_B) - R(\theta_B)$, and $R = R(\theta_A) - A(\theta_A)$, the total expected loss per compound becomes

$$[C_A p + C_B q + A\rho_B q + R(1 - \rho_A)p] + [A(\theta_A)p + R(\theta_B)q]$$

The expression in the second set of brackets defines a cost that does not depend on the test procedure used. Hence, defining the expression in the first set of brackets as the regret of the test procedure, Dunnett suggests that what one wishes to do is to minimize the expected regret per *active* compound accepted, that is, minimize

$$B = \frac{C_A p + C_B q + A\rho_B q + R(1 - \rho_A)p}{p\rho_A}$$

where $C_A$, $C_B$, $\rho_A$, and $\rho_B$ are parameters depending on the screen used.

Dunnett examines in detail the solution to the problem above for a class of procedures under the assumption that a single replication has a normal distribution centered about $\theta$ with known variance $\sigma^2$. These include single-, double-, and triple-stage screens. Although well received at the time of its publication, this model does suffer from some conceptual difficulties. In combining some of the decision-theoretic features of Hald together with an emphasis on the effort expended per active compound discovered as in the formulations of Davies (1958), a solution is obtained which is not known to be optimal for any well-formulated problem. At the same time, no guiding principles are available for helping the researcher to generalize the model to more complex distributions in which $\theta$ varies continuously.

In addition, a drawback of the foregoing procedure is the difficulty of specifying the true losses. In discussing these, Davies (1963) points out that the loss due to a "false positive is simply the cost of further testing to evaluate the compound and the resulting reduced probability of finding a drug due to unprofitable use of testing facilities." In contrast, "the cost of a false negative includes the expected cost of finding a positive, the loss resulting from the delay in finding a drug, and the opportunity afforded to competition to find a drug first, set up costs when changing from one project to the next, and so on." Some of these losses may be easily quantified, such as the loss due to further testing, but not others, such as the expected cost of finding a positive. The difficulty with the latter is that the expected cost of finding a positive is not independent of the test procedure used. In the context of acceptance sampling, this has been observed by Schlaifer (1959) and Bergman (1981).

A model that overcomes some of these difficulties has been developed by one of the present authors (Bergman, 1981). This model formalizes the notion that there are a large number of compounds that compete to be among the few that are to be accepted per time unit. Like Hald's acceptance sampling model, it is based on maximizing the expected net return. Rewards are assigned for accepting actives and inactives, respectively, and as a function of activity $\theta$ these are denoted $U(\theta)$. For a highly active compound this will represent the expected value of a marketable drug less additional testing costs, while for a relatively inactive compound it represents only these additional testing costs. The primary assumption is that there is some prior information about the range of possible values that $\theta$ may take, and that this information may be quantified as a prior distribution $g(\theta)$. In addition, it is assumed that the screening procedure entails a setup cost $C_2$ per compound submitted and an observation cost $C_1$ per replication.

The objective is to maximize the expected reward of the screen per compound accepted, and is defined by averaging over terminal rewards and the costs of testing compounds. For any given screen with operating characteristic curve $\rho(\theta)$ and average sample number $\eta(\theta)$, the expected reward may be written

$$V(\rho, n) = \frac{\int [U(\theta)\rho(\theta) - C_2 - C_1 n(\theta)]g(\theta) \, d\theta}{\int \rho(\theta)g(\theta) \, d\theta} \qquad (4)$$

By trial and error the maximizing screen may be found.

The theory leading to this formulation may be described in terms of the simplified setting in which only one compound is to be selected for further development and only one observation may be taken on each compound. To determine the form of the optimal procedure for this problem, we consider the question of how the researcher should behave given that he or she has already tested n compounds. At that point the researcher may either select for further development one of the n compounds already tested or may test additional compounds and delay the choice. Clearly, if the researcher chooses to select one of the compounds already tested, he or she should select the most promising, the value of which will be denoted $y_n$. Also, clearly, the search should be continued only if the researcher would expect to find a more promising compound whose expected value y' is sufficiently greater than $y_n$ to offset the additional search costs incurred in finding this more promising compound. Assuming that the researcher employs an optimal search procedure, this alternative will have some net expected value, which will be denoted $k_n^*$. Consequently, at stage n the optimal choice is to accept the current best if and only if $y_n \geqslant k_n^*$.

To specify the optimal procedure, we need therefore only specify the constants $k_n^*$, n = 1,2,...,N, where N is the total number of compounds available. They are not known a priori, for they represent, at each stage, the expected return from a policy which is optimal subject to the condition that the compound selected is not one of the first n tested. It may be shown (see Bergman, 1981) that $k_n^*$ takes a value k* which is independent of n for n < N. Thus the optimal procedure is one of selecting the first compound whose value is greater than k*. If there are a large number of compounds competing to be the one selected, this also implies that a compound which is not immediately selected after testing is, with high probability, rejected forever.

The foregoing rule is interesting, for it defines in a new way the cost of rejecting a compound of activity θ. This is simply the cost of the optimal search procedure for finding the best alternative. Thus we need never ask ourselves, for example, the cost of rejecting a compound that will cure a given type of cancer, as has been suggested by Colton. This is readily seen by considering the case where every compound can cure this cancer. Under these circumstances the cost of missing the first such compound is in fact less than or equal to the cost of thoroughly testing one more compound.

To find $k^*$, we may consider a reduced problem $P(k)$ for which there is just one compound to be tested, and after testing this compound the researcher may either accept it or receive a reward $k$. Denoting by $V(k)$ the expected return under an optimal policy for $P(k)$, it may be shown that $k^*$ is the unique solution of the equation $k = V(k)$ [or fixed point of the function $V(\cdot)$]. Furthermore, this solution may be obtained iteratively. Setting $k_0 = 0$ and $k_n = V(k_{n-1})$ $(n = 1, 2, \ldots)$ it may be shown that the sequence $k_0, k_1, k_2, \ldots$ converges to $k^*$ monotonically.

This solution may be generalized in several ways. First, it may be shown that if the objective is to select $r$ compounds for further development, and if the number of compounds competing for these $r$ positions is large, the optimal procedure is to apply the one compound procedure $r$ times in succession. Second, if $n$ observations are to be taken on each compound, the solution is exactly as before except that now the sampling distribution reflects the information obtained through $n$ observations rather than just one. This implies that to find the optimal $n$ one just determines $k^*$ separately for each sample size $n$ and selects that $n$ which yields the highest $k^*$. Third, it may be shown that if one is permitted to determine the total number of observations that are to be taken on a compound sequentially, this problem may also be solved by considering a $P(k)$-type problem. In particular, this version of the $P(k)$ problem consists of one compound which may be accepted, tested once again, or finally rejected at each stage, with costs and rewards as in the original problem [i.e., $C_1$, $C_2$, and $U(\theta)$], except that a final reward of $k$ which is known in advance is received if the compound is finally rejected. Letting $V(k)$ denote the value of this problem under a rule $S(k)$ which is optimal for it, we have here too that the maximum expected reward for the original problem is the unique solution of the equation $V(k) = k$. The optimal procedure corresponding to $k^*$ may then be described as follows. Select a compound for testing and test it as often as it would be optimal to do so if this were the only compound in problem $P(k^*)$. Then accept the compound if it would have been accepted in $P(k^*)$. If not, select another compound and repeat the process. Continue in this way until as many compounds have been accepted as one requires for further development. That is, sample each compound according to rule $S(k^*)$ until the required number of compounds is obtained. Techniques have been developed for finding both k* and S(k*). Typically, for a given k, say $k_n$, $S(k_n)$ is computed by backward induction (see any text on dynamic programming, e.g., Raiffa 1968). Then by defining $k_{n+1} \equiv V(k_n)$, we have as before that the sequence $k_0, k_1, k_2, \ldots$ converges to k*.

The relationship between the theory outlined briefly above and the expression given in (4) may be obtained by formally solving the equation k* = V(k*) for k* when the P(k) problem is formulated in terms of a screen's OC curve $\rho$ and average sample number $\eta$. Specifically, the expected reward of such a screen for the P(k) problem is

$$V(k|\rho,\eta) = \int \{U(\theta)\rho(\theta) + k[1 - \rho(\theta)] - C_2 - C_1\eta(\theta)\}g(\theta)\,d\theta$$

$$= k - k\int \rho(\theta)g(\theta)\,d\theta + \int [U(\theta)\rho(\theta)$$

$$- C_2 - C_1\eta(\theta)]g(\theta)\,d\theta$$

Hence, solving for k in the equation $k = V(k|\rho,\eta)$ we obtain

$$k = \frac{\int [U(\theta)\rho(\theta) - C_2 - C_1\eta(\theta)]g(\theta)\,d\theta}{\int \rho(\theta)g(\theta)\,d\theta}$$

For some pair $(\hat{\rho}, \hat{\eta})$ a largest value $\hat{k}$ will be obtained. These are the operating characteristic and average sample number functions for the optimal screen.

As an illustration of the type of results one obtains for a single-stage design, we may consider the case where $U(\theta) = \theta$, $\theta$ has a normal $N(0,1)$ prior, each replication is normally distributed with mean $\theta$ and standard deviation 0.75, $C_1 = 0.001$, and $C_2 = 0.02$. From tables in Bergman (1981) one finds that the optimal number of replications is 5 and that any compound with an expected reward greater than 1.48 should be accepted.

An example will illustrate the type of stopping boundaries one obtains in the fully sequential case. We consider the situation in which there are just two types of compounds: type A and type B. Type A is relatively active and hence potentially useful, while type B is not. Let us assume that the proportion of type A among those to be tested is 0.01. In addition, assume that a replication in a screen will register active with probability 0.8 if the compound is of type A and 0.2 if it is of type B. Assume further that the setup cost, $C_2$, is 0.025; the replica-

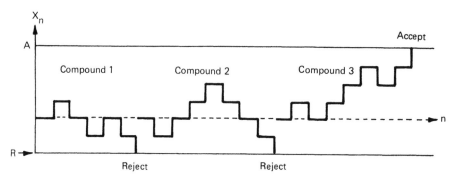

Figure 21 Typical sample path for sequential screen.

tion cost, $C_1$, is $0.005$; the reward of passing a type A compound is $25.0$; and the $(< 0)$ reward of accepting a type B is $-1.0$. It may be shown that the optimal rule for this type of problem accepts a compound if the random walk defined by $X_n = 2Y_n - n$ ($n = 1, 2, \ldots$) exceeds some value A before it falls below another value $R(<A)$, and otherwise rejects the compound. Here $Y_n$ is the number of positive responses in $n$ trials. A typical sample path is depicted in Figure 21.

The optimal policy for this problem turns out to have $A = 9$ and $R = -2$. A little less than 1% ($0.00938$) of the compounds are accepted and an average of $3.44$ replications are required per compound. The expected reward per compound accepted with this rule is $20.49$. See also Bergman (1981).

REFERENCES

Armitage, P., and Schneidermann, M. (1958), "Statistical problems in a mass screening program," Ann. N.Y. Acad. Sci., Vol. 76, pp. 896–908.

Arnow, L. E. (1970), Health in a Bottle: Searching for the Drugs That Help, Lippincott, Philadelphia.

Bergman, S. W. (1981), "Acceptance sampling: the buyer's problem," Ph.D. dissertation, Yale University.

Bergman, S. W., and Turbull, B. W. (1983), "Efficient sequential designs for destructive life testing with application to animal serial sacrifice experiments," Biometrika, Vol. 70, No. 2, pp. 305-314.

Bliss, C. J. (1967), Statistics in Biology; Statistical Methods for Research in the Natural Sciences, McGraw-Hill, New York.

Bureau of Chronic Diseases (1962), Cal. State Dept. Public Health, *Bibliography on Multiphase Screening*, Mimeo, Berkeley, Calif.

Chow, Y. S., Robbins, H., and Siegmund, D. (1971), *Great Expectations: The Theory of Optimal Stopping*, Houghton Mifflin, Boston.

Colton, T. (1963), "Optimal drug screening plans," *Biometrika*, Vol. 50, pp. 31-45.

Curnow, R. N. (1961a), "Optimal programming for varietal selection," *J. R. Statist. Soc. B*, Vol. 23, pp. 311–318.

Curnow, R. N. (1961b), "Optimal programmes for varietal selection," *J. R. Statist. Soc. B*, Vol. 23, pp. 282–292.

Curnow, R. N. (1965), "A note on Watson's paper," *Technometrics*, Vol. 7, pp. 444–446.

Davies, O. L. (1958), "The design of screening tests in the pharmaceutical industry," *Bull Int. Statist. Inst.*, Vol. 36, III, pp. 226–240.

Davies, O. L. (1959), "The design of screening tests in the pharmaceutical industry," *Bull Int. Statist. Inst.*, Vol. 36, I, pp. 85–86.

Davies, O. L. (1963), "The designer of screening test," *Technometrics*, Vol. 5, pp. 481–490.

Davies, O. L. (1964), "Screening for improved mutants in antibiotic research," *Biometrics*, Vol. 20, pp. 576–591.

Dudewicz, E. (1980), "Ranking (ordering) and selection: an overview of how to select the best," *Technometrics*, Vol. 22, No. 1.

Dunnett, C. W. (1960), "The statistical theory of drug screening and other problems connected with the screening, testing and selection of drugs," D.Sc. thesis, University of Aberdeen.

Dunnett, C. W. (1961), "The statistical theory of drug screening," *Quantitative Methods in Pharmacology*, North-Holland, Amsterdam, pp. 212–231.

Dunnett, C. W. (1972), in *Statistics: A Guide to the Unknown*, J. Tanur, ed., Holden-Day, San Francisco, pp. 26–37.

Elandt, R. C. (1963), Optimal and sufficient allocation of multiple varietal experiments, *Biometrics*, Vol. 19, pp. 615–628.

Federer, W. T. (1963), "Procedures and designs useful for screening material in selection and allocation with a bibliography," *Biometrics*, Vol. 19, pp. 553–587.

Finney, D. J. (1957), "The consequences of selection for a variate subject to errors of measurement," *Rev. Inst. Int. Statist.*, Vol. 24, pp. 22–29.

Finney, D. J. (1958), "Statistical problems of plant selection," *Bull Inst. Int. Statist.*, Vol. 36, pp. 242--268.

Finney, D. J. (1959), "The economic efficiency of experimentation," *Publ. Math. Inst. Hung. Acad. Sci.*, Vol. 4, pp. 203--224.

Finney, D. J. (1960a), "A simple example of the external economy of varietal selection," *Bull Int. Statist. Inst.*, Vol. 37, III, pp. 91--106.

Finney, D. J. (1960b), *An Introduction to the Theory of Experimental Design*, University of Chicago Press, Chicago.

Finney, D. J. (1966), "An experimental study of certain screening processes," *J. R. Statist. Soc. B*, Vol. 28.

Freidlina, V. L. (1975), "On the problem of design for screening experiments," *Theory Prob. Appl.*, Vol. 20, pp. 102--115.

Gart, J. J. (1961), "Modification of a sequential procedure for radiation protection studies with mice," *Radiat. Res.*, Vol. 15, pp. 616--624.

Goldstein, (1975), "Selection of variates for the two group multinominal classification problem," *J. Am Statist. Assoc.*, Vol. 70, pp. 776--781.

Grimshaw, J. J., and D'Arcy, P. F. (1961), "Some problems arising in drug standardization and the selective screening of compounds of potential pharmacological interest," in *Quantitative Methods in Pharmacology*, H. DeJonge, ed., North-Holland, Amsterdam.

Gross, A. J., and Mantel, N. (1967), "The effective use of both positive and negative controls in screening experiments," *Biometrics*, Vol. 23, pp. 285--295.

Hald, A. (1960), "The compound hypergeometric distribution and a system of single sampling inspection plans based on prior distribution and costs," *Technometrics*, Vol. 2, pp. 275--340.

King, E. P. (1963), "A statistical design for drug screening," *Biometrics*, Vol. 19, pp. 429--440.

King, E. P. (1964), "Optimal replication in sequential drug screening," *Biometrika*, Vol. 51, pp. 1--10.

Kleyin, J. P. (1975), "Screening designs for poly-factor experimentation," *Technometrics*, Vol. 17, pp. 487--493.

Lee-Jen Wei (1977), "Sequential searching schemes for an optimal dosage," *Aust. J. Statist.*, Vol. 19, pp. 163--171.

Li, C. H. (1962), "A sequential method for screening experimental variables," *J. Am. Statist. Assoc.*, Vol. 57, pp. 455--477.

Mantel, N. (1961), "Principles of chemotherapeutic screening," *Proc. 4th Berkeley Symp. Math. Statist. Prob.*, Vol. 4, pp. 293–306.

Owen, D. B., and Su, Y. H. (1977), "Screening based on normal variables," *Technometrics*, Vol. 19, pp. 65–68.

Raiffa, H. (1968), *Decision Analysis; Introductory Lectures on Choices Under Uncertainty*, Addison-Wesley, Reading, Mass.

Rosenberry, T. D., and Gehan, E. A. (1964), "Operating characteristic curves and accept-reject rules for two and three-stage screening procedures," *Biometrics*, Vol. 20, No. 1.

Schlaifer, R. (1959), *Probability and Statistics for Business Decisions*, McGraw-Hill, New York.

Schultz, J. R., Nichol, F. R., Elffring, G. L., and Weed, S. D. (1973), "Multiple-stage procedures for drug screening," *Biometrics*, June, pp. 293–300.

Scott, A. (1969), "A note on an allocation problem," *J. R. Statist. Soc. B*, Vol. 31, pp. 119–122.

Watson, G. S. (1961), "A study of the group screening method," *Technometrics*, Vol. 3, pp. 371–388.

Wetherill, G. B. (1975), "A review of Acceptance Sampling Schemes with emphasis on the Economic Aspect," *Int. Statist. Rev.*, Vol. 43, No. 2, pp. 191–210.

Wilson, E. B., and Burgess, A. R. (1971), "Multiple sampling plans viewed as finite Markov chains," *Technometrics*, Vol. 13, pp. 371–383.

Young, J. C. (1972), "An investigation of procedures for multiple-stage selection for a variate subject to errors of measurement," *Biometrika*, Vol. 59, pp. 323–334.

Young, J. C. (1974), "Multiple-stage screening from non-normal populations," *Biometrika*, Vol. 61, No. 1, pp. 155–163.

# 3

## ALLOCATION OF RESOURCES BETWEEN
## RELATED SCREENS

### 3.1. INTRODUCTION

In Chapter 2 methods were described for improving the design of
a single preliminary screen by the use of replications. The principles discussed there may also be applied to subsequent screens,
and each may be optimized in isolation. However, it is clear that
optimizing each individual screen is not equivalent to optimizing
the total screening effort inherent in discovering a marketable
drug. In this chapter we discuss our third major area of focus,
allocation of resources between different but related screens.
Under this heading we include (1) selection of screens to be included in the total screening effort, (2) sequencing of a collection
of given screens, and (3) resource allocation between these
screens.

In spite of their importance, little attention has been given
to any of these areas either in the statistical literature or in the
laboratories we have visited. Consequently, we have attempted
to devise some models that will answer pertinent questions related
to them.

### 3.2. RESOURCE ALLOCATION BETWEEN OR-
### DERED SCREENS

We begin with problem (3)—how resources should be allocated to
a collection of related screens sequenced in a fixed preassigned

order. This will include subarea (1) as well when we have the
convention that selection of a screen to be included in a sequence
of screens is equivalent to allocating resources to that screen.

In essence, our understanding of the resource allocation
problem is the following. Each screen in a sequence of screens
serves to give information about some unknown parameter $\theta*$ that
is of interest. For example, $\theta*$ may be the activity of a compound
in safely reducing hypertension in humans. For both ethical and
economic reasons we do not immediately test unknown compounds
on humans. Instead, the compound is first submitted to a sequence of proxy screens that may include, for example, tests in
vitro, tests on mice, and tests on monkeys. What we learn at
each screen is the hypertension-related activity of the compound
for the proxy subject of that screen, be it a culture, a mouse,
or a monkey.

Formally, we may model the relationship between the activities
of a compound on two given screens as one of correlated random
variables. Thus, if $\theta_i$ is the true activity of the compound on
screen i, this defines a conditional density $g(\theta_j|\theta_i)$ for its activity
$\theta_j$ on screen j, as well as a conditional density $g(\theta*|\theta_i)$ for its
activity $\theta*$ in humans. However, at no given screen i do we observe $\theta_i$ directly, only a random variable $X_i$ whose distribution conditional on $\theta_i$ is assumed known. Thus a single observation on the
ith screen provides only partial information about $\theta_i$, which in turn
provides only partial information about $\theta*$. As in Chapter 2, we
replicate at each screen in to learn $\theta_i$ with increasing precision.
However, as our ultimate goal is to learn about $\theta*$, there is a point
at which it becomes more profitable to move to the next screen
and take observations there rather than continue with replications
at the current screen.

To get a handle on this problem, we have extended the decision-theoretic model discussed at the end of Chapter 2. It permits such an extension since it is defined in terms of rewards,
and hence is able to balance the values and costs of the information that may be gained at the different screens.

As in that model, let $U(\theta*)$ be the reward from a compound
shown to possess activity $\theta*$ for humans. If we were to consider
only one screen in isolation, say a testing program administering
compounds to consenting patients, the model would be as before.
We may denote the setup cost per compound by $C_2^*$ and the cost
per observation by $C_1^*$. Assuming a known prior frequency $g(\theta*)$
for the activities of the compounds submitted to this screen, then
within the class of screening procedures we are willing to con-

sider (single stage, multistage, or fully sequential screen), we would find that screen whose operating characteristic curve $\rho(\theta^*)$ and average sample number function $\eta(\theta^*)$ (see Section 2.6 for definitions) maximizes the ratio

$$\frac{\int [U(\theta^*)\rho(\theta^*) - C_2^* - C_1^*\eta(\theta^*)]g(\theta^*)d\theta^*}{\int \rho(\theta^*)g(\theta^*)d\theta^*}$$

Although this would be prohibitively expensive, as well as un-ethical, we may assume initially that the corresponding optimal screen thus obtained is the only screen and that all compounds are tested on it. Then $g(\theta^*)$ would be equal to the prior for $\theta^*$ as found among all untested compounds.

Let us now consider what happens when an additional screen, say one on monkeys, is added. There are two simultaneous effects. On the positive side, interposing this screen has the effect of filtering out some of the relative inactive compounds, thus altering the prior $g(\theta^*)$ of the compounds submitted to the clinical trial screen. On the negative side, $C_2^*$ is increased, reflecting the costs expended in the monkey screen. Exactly how $g(\theta^*)$ and $C_2^*$ are altered will depend on the operating characteristic curve $\rho(\theta)$ and average sample number function $\eta(\theta)$ of the preliminary screen. By Bayes' theorem, the new prior for the clinical screen becomes

$$\tilde{g}(\theta^*) = g(\theta^*) \frac{\int \rho(\theta)g(\theta|\theta^*)d\theta}{\int \rho(\theta)g(\theta)d\theta}$$

and the new setup cost is

$$\tilde{C}_2^* = C_2^* + \frac{C_2 + C_1\int \eta(\theta)g(\theta)d\theta}{\int \rho(\theta)g(\theta)d\theta}$$

where $C_2$ and $C_1$ are the setup cost and cost per observation for the preliminary screen. Consequently, the expected reward $V((\rho,\eta),(\rho^*,\eta^*))$ associated with using these two screens in conjunction is

$$\int [U(\theta^*)\rho(\theta^*) - \tilde{C}_2^* - C_1^*\eta(\theta^*)]\tilde{g}(\theta^*)d\theta^*$$

or, equivalently,

$$\frac{\iint\{[U(\theta^*)\rho(\theta^*) - C_2^* - C_1^*n(\theta^*)]\rho(\theta) - C_2 - C_1 n(\theta)\}g(\theta,\theta^*)\,d\theta\,d\theta^*}{\iint\rho(\theta^*)\rho(\theta)g(\theta^*,\theta)\,d\theta\,d\theta^*}$$

where $g(\theta^*,\theta)$ is the joint distribution of $(\theta^*,\theta)$ obtained from the relationship $g(\theta^*,\theta) = g(\theta^*|\theta)g(\theta)$.

One may also define an operating characteristic curve for this sequence of screens as the probability that a compound of activity $\theta^*$, the variable of interest, will successfully pass through both screens. Denoting the OC curve by $\tilde{\rho}(\theta^*)$, we have

$$\tilde{\rho}(\theta^*) = \rho(\theta^*)\int\rho(\theta)g(\theta|\theta^*)\,d\theta$$

In contrast, it does not make sense to define an average sample number function solely in terms of the number of replications required by the two screens, since the cost of a replication on the final screen is usually much more expensive than a replication on the preliminary screen. Rather, these replications should be weighted by their relative costs. Doing so, we obtain an average testing cost function $\tilde{C}(\theta^*)$ defined by

$$\tilde{C}(\theta^*) = [C_2^* + C_1^*n(\theta^*)]\int\rho(\theta)g(\theta|\theta^*)\,d\theta$$

$$+ \int[C_2 + C_1 n(\theta)]g(\theta|\theta^*)\,d\theta$$

The curves $\tilde{\rho}(\theta^*)$ and $\tilde{C}(\theta^*)$ may now be used to compare different designs which vary not only with respect to the design of the individual screens but also with respect to the number of screens within the sequence.

To find the best screen sequence we now maximize $V((\rho,n),(\rho^*,n^*))$ in $(\rho,\nu)$ and $(\rho^*,n^*)$ simultaneously. This may be done by a trial-and-error search procedure or by more sophisticated procedures based on fixed points (see Bergman, 1981).

From an administrative point of view, it is interesting to consider the relationship between this joint maximization problem and the individual maximization problems faced by two research groups, each operating one of these screens and each trying to optimize its own screen given the inputs or outputs of the other screen. For example, we may consider the following sequence of events. First the operators of the final screen optimize their screen with respect to their inputs as defined by the current preliminary screen. Then the operators of the preliminary screen optimize their own screen with respect to the new set of parameters now characterizing the final screen. This, in turn, leads the operators of the final screen to reoptimize their screen. And so on. In general, this sequence of suboptimal solutions will not converge to the optimal solution of the two-screen problem. However, it will lead to a local optimal solution of the two-screen problem and hence, if one begins in the vicinity of the global optimum, the local optimum obtained and the true global optimum are likely to coincide.

The question that remains is exactly how these individual maximization problems should be defined so as to achieve this convergence to at least a local optimum of the two-screen problem. Since

$$\max_{(\rho*,\eta*)} V((\rho*,\eta*),(\rho,\eta))$$

$$= \max_{(\rho*,\eta*)} \frac{\int [U(\theta*)\rho(\theta*) - C_2^* - \tilde{C}_1^*\eta(\theta*)]\tilde{g}(\theta*) \; d\theta*}{\int \rho(\theta*)\tilde{g}(\theta*) \; d\theta*}$$

where $\tilde{g}(\theta*)$ and $\tilde{C}_2^*$ are defined as before, the operators of the final screen should behave as if they are trying to optimize a single screen with reward $U(\theta*)$, prior $\tilde{g}(\theta*)$, setup cost $\tilde{C}_2^*$, and sampling cost $C_1^*$. Note, however, that a rather different formulation is obtained for the preliminary screen. Setting

$$U(\theta) = \int [U(\theta*)\rho(\theta*) - C_2^* - C_1^*\eta(\theta*)]g(\theta*|\theta) \; d\theta*$$

and

$$h(\theta) = \int \rho(\theta*)g(\theta*|\theta) \; d\theta*$$

we obtain

$$\max_{(\rho,\eta)} V((\rho*,\eta*),(\rho,\eta))$$

$$= \max_{(\rho,\eta)} \frac{\int [U(\theta)\rho(\theta) - C_2 - C_1\eta(\theta)]g(\theta) \; d\theta}{\int h(\theta)\rho(\theta)g(\theta) \; d\theta}$$

Thus the preliminary screen operators acting in isolation should be trying to optimize a slightly different problem: namely, one characterized by reward $U(\theta)$, prior $g(\theta)$, setup cost $C_2$, sampling cost $C_1$, and a new factor $h(\theta)$, which defines the probability that a compound passed will in fact also pass the final screen.

The result above shows that for the situation where at most one compound for a particular disease will be marketed by the firm, the optimization problem defined for the preliminary screen in a sequence of screens is not the same as the optimization problem for a single screen.

To illustrate the types of results one may obtain, a program has been written to analyze the allocation problem for two or three screens under the following assumptions. First, a compound is either inherently active or inactive for the biological system characterizing a screen. Second, each observation on a compound within a screen may take one of two values: positive or negative. Third, there is a known joint probability matrix defining the simultaneous probability of a compound being inherently active or inactive at each of the screens. Fourth, the observational error is symmetric at each screen; that is, the probability that a replication on a screen will yield a false negative is equal to the probability that a replication will yield a false positive. This assumption is restrictive but makes the analysis relatively easy. Fifth, for each screen there is specified a setup cost per compound submitted and a sampling cost per observation. Sixth, only a small proportion of all compounds will be marketed. The value of a compound, as before, depends only on its inherent activity for the last screen, that is, its activity in humans.

Under these assumptions it may be shown that the optimal sequence of screens is of the following form. Let $X_{ij}$ be either 0 or 1, depending on whether the jth replication of a compound on the ith screen is negative or positive, and set $Y_{im} = 2(\Sigma_{j=1}^{m} X_{ij}) - m$. Then there exist boundaries $a_i$ and $r_i$ for the ith screen such that one continues replicating whenever $r_i < Y_{im} < a_i$, and one finally stops and either accepts the compound for testing at the $(i + 1)$th screen (markets the compound if there are only i screens), or rejects the compound, at the first m for which $Y_{im} \leq r_i$ or $Y_{im} \geq a_i$, respectively.

Tables 5 through 11 present the output from this program in the case of three screens. The reward for marketing a compound inherently active or inactive on the final screen is specified in the UTILITY VECTOR. In Table 5 these quantities are 25.0 and $-1.0$, respectively. The probability of a true positive (negative) for each of the three screens is given in the THETA VECTOR.

Hence, for the example displayed on Table 5, the probability of a
false positive (negative) on a replication in any of the screens is
0.20. The testing costs of the ith screen are given in the ith row
of the COST MATRIX. The first entry defines the cost per observation, and the second the setup cost per compound. The
(i,j,k)th entry of the PROBABILITY MATRIX defines the probability that a compound will simultaneously be of inherent activity
i for the first screen, j for the second, and k for the third,
where, for example, i is either 1 (inactive) or 2 (active). Thus,
in Table 5 we see that the probability of a compound being active
on the first screen, and inactive on the second and third, is
0.0706. From this probability matrix one may derive (1) the
ACTIVE PROBABILITY VECTOR, which gives the probability of
a compound picked at random being active for the first, second,
and third screens, respectively; (2) the 2ND STAGE CONDITION-
AL ACTIVE PROBABILITY VECTOR, which gives the probability
that an inactive or active on the first screen will be active for the
second screen; and (3) the 3RD STAGE CONDITIONAL ACTIVE
PROBABILITY MATRIX, which gives the probability for each of
the four combinations of active and inactive for the first two
screens leading to activity on the third screen. The program also
permits the user to specify which of the screens the researcher is
permitted to use. This is set in the parameter SCREEN, where 1
in the ith entry means that the ith screen may be used, and 0 that
it may not.

For the specification given in Table 5, the optimal solution is
to set $a_1 = 4$, $r_1 = -1$, $a_2 = 2$, $r_2 = -2$, $a_3 = 7$, and $r_3 = -2$.
This yields an expected reward of 11.87. The prior effective
distribution $\tilde{g}(\theta)$ for the ith screen is given as the ith entry of
POM. There we see that the prior probability of an active compound at the first screen is 0.100, the same as for the original
population of untested compounds. However, after filtering by
the first screen the effective prior probability of an active compound at the second screen is raised from the original value of
0.05 to 0.2816. Similarly, after filtering by the first and second
screens, the effective prior probability of an active compound in
humans at the third screen has been raised from 0.002 to 0.0332.
The effective setup cost of the compounds supplied to each of the
screens under the optimal rule is given in row C2M. Thus this
cost for the second screen has been raised from 0.01 to 0.0735,
reflecting the cost of filtering by the first screen. Finally, the
ith entry of the rows EM and QM specify, respectively, the average number of replications that a compound submitted to the ith
screen by the previous screens will experience at this screen, and
the probability that it will pass this screen.

TABLE 5  Three-Stage Fully Sequential Symmetric Two-Point Procedure

INPUT PARAMETERS ARE:

SCREEN   1   1   1

UTILITY VECTOR:        25.00000        −1.00000

THETA VECTOR:          0.8000000       0.8000000       0.8000000

COST MATRIX:

| | | |
|---|---|---|
| 0.0010000 | 0.0030000 | |
| 0.0050000 | 0.0100000 | |
| 0.0200000 | 0.0500000 | |

ACTIVE PROBABILITY VECTOR:        0.1000000        0.05000000        0.0200000

2ND STAGE CONDITIONAL ACTIVE PROBABILITY VECTOR:

0.02325558        0.2906977

3RD STAGE CONDITIONAL ACTIVE PROBABILITY MATRIX:

| 0.0001896 | 0.0189606 |
|---|---|
| 0.0047352 | 0.0378813 |

PROBABILITY MATRIX:

|  | SCREEN 3 INACTIVE | SCREEN 3 ACTIVE |
|---|---|---|
| SCREEN 1 INACTIVE | 0.8789033    0.0205338 | 0.001665    0.0003964 |
| SCREEN 1 ACTIVE | 0.0705944    0.0279686 | 0.0003359    0.0011012 |

THE OPTIMAL SOLUTIONS ARE:

EXPECTED NET RETURN    11.86588

| IAM: | 4 | 2 | 7 |
|---|---|---|---|
| IRM: | $-1$ | $-2$ | $-2$ |
| POM: | 0.1000000 | 0.2816147 | 0.0332487 |
| QM: | 0.0777126 | 0.3073071 | 0.0312261 |
| EM: | 1.9369501 | 2.9411765 | 3.5784070 |
| C2M: | 0.00300 | 0.07353 | 0.33712 |

TABLE 6  As Table 5 Using Screen 3 Only

INPUT PARAMETERS ARE:

SCREEN:     0    0    1

UTILITY VECTOR:     25.00000     −1.00000

THETA VECTOR:     0.8000000     0.8000000     0.8000000

COST MATRIX:

|  |  |  |
|---|---|---|
| 0.0010000 | 0.0030000 | |
| 0.0050000 | 0.0100000 | |
| 0.0200000 | 0.0500000 | |

ACTIVE PROBABILITY VECTOR:     0.1000000     0.050000     0.0020000

2ND STAGE CONDITIONAL ACTIVE PROBABILITY VECTOR:

0.0232558     0.2906977

3RD STAGE CONDITIONAL ACTIVE PROBABILITY MATRIX:

| 0.0001894 | 0.0189406 |
|-----------|-----------|
| 0.0047352 | 0.0378813 |

PROBABILITY MATRIX:

|                   | SCREEN 3 INACTIVE | | SCREEN 3 ACTIVE | |
|-------------------|-----------|-----------|-----------|-----------|
| SCREEN 1 INACTIVE | 0.8789033 | 0.0205338 | 0.0001665 | 0.0003964 |
| SCREEN 1 ACTIVE   | 0.0705944 | 0.0279686 | 0.0003359 | 0.0011012 |

THE OPTIMAL SOLUTIONS ARE:

| EXPECTED NET RETURN | | | 0.94800 |
|-----------|-----------|-----------|-----------|
| IAM: | 0 | 0 | 0 |
| IRM: | $-1$ | $-1$ | $-1$ |
| POM: | 0.1000000 | 0.0500000 | 0.0020000 |
| QM: | 1.0000000 | 1.0000000 | 1.0000000 |
| EM: | 0.0000000 | 0.0000000 | 0.0000000 |
| C2M: | 0.00300 | 0.01000 | 0.05000 |

**TABLE 7** As Table 5 Using Screens 2 and 3 Only

INPUT PARAMETERS ARE:

SCREEN:　　　0　　1　　1

UTILITY VECTOR:　　25.00000　　−1.00000

THETA VECTOR:　　0.8000000　　0.8000000　　0.8000000

COST MATRIX:

| | |
|---|---|
| 0.0010000 | 0.0030000 |
| 0.0050000 | 0.0100000 |
| 0.0200000 | 0.0500000 |

ACTIVE PROBABILITY VECTOR:　　0.1000000　　0.05000000　　0.0020000

2ND STAGE CONDITIONAL ACTIVE PROBABILITY VECTOR:

0.0232558　　0.2906977

3RD STAGE CONDITIONAL ACTIVE PROBABILITY MATRIX:

| | |
|---|---|
| 0.0001894 | 0.01894606 |
| 0.0047352 | 0.0378813 |

PROBABILITY MATRIX:

| | SCREEN 3 INACTIVE | | SCREEN 3 ACTIVE | |
|---|---|---|---|---|
| SCREEN 1 INACTIVE | 0.0205338 | 0.0205338 | 0.0001665 | 0.0003964 |
| SCREEN 1 ACTIVE | 0.0705944 | 0.0279686 | 0.0003359 | 0.0011012 |

THE OPTIMAL SOLUTIONS ARE:

EXPECTED NET RETURN: 2.43894

| | | | |
|---|---|---|---|
| IAM: | 0 | 4 | 6 |
| IRM: | $-1$ | $-1$ | $-2$ |
| POM: | 0.1000000 | 0.05000000 | 0.0279197 |
| QM: | 1.0000000 | 0.0403226 | 0.0263976 |
| EM: | 0.0000000 | 1.7895894 | 3.4932371 |
| C2M: | 0.00300 | 0.01000 | 0.51991 |

TABLE 8  As Table 5 Using Screens 1 and 3 Only

INPUT PARAMETERS ARE:

SCREEN:        1        0        1

UTILITY VECTOR:        25.0000        − 1.00000

THETA VECTOR:        0.8000000        0.8000000        0.8000000

COST MATRIX:

0.0010000        0.0030000

0.0050000        0.0100000

0.0200000        0.0500000

ACTIVE PROBABILITY VECTOR:        0.1000000        0.0500000        0.0020000

2ND STAGE CONDITIONAL ACTIVE PROBABILITY VECTOR:

0.0232558        0.2906977

3RD STAGE CONDITIONAL ACTIVE PROBABILITY MATRIX:

| 0.0001894 | 0.0189406 |
| 0.0047352 | 0.0378813 |

PROBABILITY MATRIX:

| | SCREENS INACTIVE | | SCREENS ACTIVE | |
|---|---|---|---|---|
| SCREEN 1 INACTIVE | 0.8789033 | 0.0205-38 | 0.0001665 | 0.0003964 |
| SCREEN 1 ACTIVE | 0.0705944 | 0.0279686 | 0.0003359 | 0.0011012 |

THE OPTIMAL SOLUTIONS ARE:

EXPECTED NET RETURN        11.04282

| IAM: | 5 | 0 | 8 |
| IRM: | − 1 | − 1 | − 2 |
| POM: | 0.1000000 | 0.2883676 | 0.0142509 |
| QM: | 0.0756777 | 1.0000000 | 0.0133743 |
| EM: | 2.0769231 | 0.0000000 | 3.4607629 |
| C2M: | 0.00300 | 0.07709 | 0.11709 |

Table 9 As Table 5 with Increased Activity in Humans

INPUT PARAMETERS ARE:

SCREEN:          1          1          1

UTILITY VECTOR:          25.00000          −1.00000

THETA VECTOR:          0.8000000          0.8000000          0.8000000

COST MATRIX:

|            |            |
|------------|------------|
| 0.0010000  | 0.0030000  |
| 0.0050000  | 0.0100000  |
| 0.0200000  | 0.0500000  |

ACTIVE PROBABILITY VECTOR:          0.1000000          0.05000000          0.0200000

2ND STAGE CONDITIONAL ACTIVE PROBABILITY VECTOR:

0.0232558          0.2906977

3RD STAGE CONDITIONAL ACTIVE PROBABILITY MATRIX:

| 0.0018941 | 0.1894065 |
|-----------|-----------|
| 0.0473516 | 0.3788129 |

PROBABILITY MATRIX

|                   | SCREEN 3 INACTIVE |           | SCREEN 3 ACTIVE |           |
|-------------------|-------------------|-----------|-----------------|-----------|
| SCREEN 1 INACTIVE | 0.8774048         | 0.0169659 | 0.0016650       | 0.0039643 |
| SCREEN 1 ACTIVE   | 0.0675716         | 0.0180578 | 0.0033587       | 0.0110120 |

THE OPTIMAL SOLUTIONS ARE:

EXPECTED NET RETURN    23.56825

| IAM: | 4         | 2         | 6         |
|------|-----------|-----------|-----------|
| IRM: | − 1       | 2         | − 2       |
| POM: | 0.1000000 | 0.2816147 | 0.3344867 |
| QM:  | 0.0777126 | 0.3073071 | 0.3118638 |
| EM:  | 1.9369501 | 2.9411765 | 5.2708653 |
| C2M: | 0.00300   | 0.07353   | 0.33712   |

Table 10  As Table 9 with Increased Screen 2 Costs

INPUT PARAMETERS ARE:

SCREEN:     1      1      1

THETA VECTOR:     0.8000000     0.8000000     0.8000000

COST MATRIX:

|           |           |
|-----------|-----------|
| 0.0010000 | 0.0030000 |
| 0.0100000 | 0.0250000 |
| 0.0200000 | 0.0500000 |

ACTIVE PROBABILITY VECTOR:     0.1000000     0.0500000     0.0200000

2ND STAGE CONDITIONAL ACTIVE PROBABILITY VECTOR:

0.0232558     0.2906977

3RD STAGE CONDITIONAL ACTIVE PROBABILITY MATRIX:

| | |
|-----------|-----------|
| 0.0018941 | 0.1894065 |
| 0.0473516 | 0.3788129 |

PROBABILITY MATRIX:

|  | SCREEN 3 INACTIVE | | SCREEN 3 ACTIVE | |
|---|---|---|---|---|
| SCREEN 1 INACTIVE | 0.8774048 | 0.0169859 | 0.0016659 | 0.0039643 |
| SCREEN 1 ACTIVE | 0.0675716 | 0.0180578 | 0.0033587 | 0.0110120 |

THE OPTIMAL SOLUTIONS ARE:

EXPECTED NET RETURN    23.46466

|  |  |  |  |
|---|---|---|---|
| IAM: | 5 | 0 | 6 |
| IRM: | $-1$ | $-2$ | $-2$ |
| POM: | 0.1000000 | 0.2883676 | 0.1425091 |
| QM: | 0.0756777 | 1.0000000 | 0.1338006 |
| EM: | 2.0769231 | 0.0000000 | 4.1620467 |
| C2M: | 0.00300 | 0.09209 | 0.11709 |

**Table 11** As Table 9 with Increased Reward for Activity in Humans

INPUT PARAMETERS ARE:

SCREEN:    1    1    1

UTILITY VECTOR:    100.00000    −1.00000

THETA VECTOR:    0.8000000    0.8000000    0.8000000

COST MATRIX:

|            |            |
|------------|------------|
| 0.0010000  | 0.0030000  |
| 0.0050000  | 0.0100000  |
| 0.0200000  | 0.0500000  |

ACTIVE PROBABILITY VECTOR:    0.1000000    0.0500000    0.0200000

2ND STAGE CONDITIONAL ACTIVE PROBABILITY VECTOR:

0.0232558    0.2906977

3RD STAGE CONDITIONAL ACTIVE PROBABILITY VECTOR:

0.0018941   0.1890065
0.0473516   0.3788129

PROBABILITY MATRIX:

|  | SCREEN 3 INACTIVE | | SCREEN 3 ACTIVE | |
|---|---|---|---|---|
| SCREEN 1 INACTIVE | 0.8774068 | 0.0169659 | 0.0016650 | 0.0039643 |
| SCREEN 1 ACTIVE | 0.0675716 | 0.0180578 | 0.0033587 | 0.0110120 |

THE OPTIMAL SOLUTIONS ARE:

EXPECTED NET RETURN    98.53466

| | | | |
|---|---|---|---|
| IAM: | 4 | 2 | 7 |
| IRM: | $-1$ | $-2$ | $-2$ |
| POM: | 0.1000000 | 0.2816147 | 0.3324867 |
| QM: | 0.0777126 | 0.3073071 | 0.3117457 |
| EM: | 1.9369501 | 2.9411765 | 5.7917944 |
| C2M: | 0.00300 | 0.07353 | 0.33712 |

We may now change the number of screens permitted to see how this alters the solution. First, in Table 6 only the last screen is permitted to be used. When optimized the expected reward is $-0.94800$; hence it is not profitable to use a single screen. Second, in Table 7 only the last two screens may be used. We note that the acceptance and rejection rules for the second screen are both shifted upward, the former from 2 to 4 and the latter from $-2$ to $-1$. In contrast, the final screen design remains unchanged even though the effective prior is now 0.0279 rather than 0.0332. We also note that the value has fallen dramatically, from 11.87 to 2.439. In Table 8, the first and third screens are permitted but not the second. When this configuration is optimized, an expected reward of 11.04 is obtained, 93% of the value for the best three-stage design. Furthermore, the boundaries of the first and third screens are shifted, becoming slightly more stringent. From these results we observe the usefulness of an inexpensive preliminary screen in achieving a high expected reward.

In the remaining tables the parameters defining the problem are altered to see how this affects the optimal solution. In Table 9 the concentration of actives with respect to humans is increased tenfold, from 0.002 to 0.02. While this increases the expected reward for the three-screen configuration from 11.86588 to 23.56825, the optimal solution is barely altered, the third-stage acceptance boundary moving down from 7 to 6. In Table 10, the second-screen cost parameters are approximately doubled, the setup cost and sampling cost moving from 0.01 to 0.025 and from 0.005 to 0.01, respectively, while the other parameters of Table 9 are held constant. The optimal solution to this problem now calls for the elimination of the second screen, and otherwise slightly alters only the first-stage screen. In Table 11, the reward for discovering an active with respect to humans is increased fourfold, from 25.0 to 100.0. The optimal solution from that found in Table 5 is barely changed, only the acceptance boundary of the final stage changing from 6 to 7. This new solution is identical to the one found in Table 5.

From the evidence, limited as it is, it appears that the model is most sensitive to those parameters about which one is apt to know the most (i.e., the cost of operating the screen) and is least sensitive to those about which one is apt to know the least (i.e., the ultimate reward and the prior distribution). This is what one would have hoped for. The theoretical background to this model and its application under more realistic assumptions are discussed in Bergman (1985).

## 3.3. OPTIMAL SEQUENCING OF SCREENS

In Section 3.2 we assumed that the order of the collection of screens under consideration was predetermined. In this section we consider the possibility of altering the sequencing to improve the total design, under the assumption that the design of each screen is fixed beforehand.

For some sets of screens there is already a natural order in which the screens should be placed. For example, in testing for activity it is often self-evidently correct to plan an in vitro screen before a mouse screen, and a mouse screen before a monkey screen. Less self-evident is the proper interspersing of preclinical activity and toxicity screens. One may argue that it is better to discover that an interesting compound is toxic, and therefore unusable, before expending considerable resources on determining with precision its degree of activity in costly test animals such as monkeys. Or conversely, one might argue that we should be sure that it is truly active before trying to determine potential drawbacks such as toxicity.

This problem (Bergman, 1981) and other more general problems concerning sequencing (see, e.g., Gittins, 1980, 1982), may be approached by the methods of dynamic programming. These methods will be illustrated by solving a model for the activity-toxicity sequencing problem.

Suppose that each compound must pass through two series of screens before being considered for final development into a marketable drug. The first series consists of a sequence of activity screens, labeled $a_1, a_2, \ldots, a_m$, which increase in complexity and are performed in the order given. Similarly, the second series, labeled $b_1, \ldots, b_n$, are the toxicity screens, beginning with a simple animal screen and proceeding through increasingly complex and expensive biological systems. We are interested in determining the optimal sequence through which the compounds should pass, given that screen $a_i$ must be performed before $a_{i+1}$, and screen $b_j$ must be performed before screen $b_{j+1}$.

If a compound fails at any of the screens, it is rejected and is not submitted to any other screen. The likelihood of failure will depend on how well it performed on the previous screens, and which screens these are will depend on the sequence chosen. Let us denote, therefore, by $P_{ij}^a$ the probability of a compound successfully passing screen $a_i$, given that it has successfully passed through screens $a_1, \ldots, a_{i-1}$ and $b_1, \ldots, b_{j-1}$. Similarly, the probability of passing successfully through $b_j$, conditional on the same previous history, is denoted $P_{ij}^b$.

The quantity to be minimized is the expected cost of determining whether a compound is potentially marketable. Since a compound is considered potentially marketable if and only if it has successfully passed through all the screens, the only question of interest is how cheaply we can determine whether a compound fails at at least one screen. The actual screening costs at each step may depend on the sequence chosen. Hence let $C_{ij}^a$ be the conditional cost of screen $a_i$ given that the compound has successfully passed through screens $a_1, \ldots, a_{i-1}$ and $b_1, \ldots, b_{j-1}$. A similar notation holds for passing through $b_j$.

It is easily seen that there are $\binom{m+n}{n}$ (i.e., the number of ways of selecting n objects from a total of m + n objects) possible sequences of screens. Let $\gamma_i$ (i = 1, 2, ..., m + n) denote the ith screen in one such sequence. The expected cost per compound of operating this sequence of screens is

$$\sum_{i=1}^{m+n} C_{\gamma i} \left( \prod_{j=0}^{i-1} P_{\gamma j} \right)$$

where $P_{\gamma 0} = 1$ and $P_{\gamma i}$ and $C_{\gamma i}$ are, respectively, the probability and cost of passing successfully through screen i, given that the compound has already passed through screens $\gamma_1, \ldots, \gamma_{i-1}$. Thus we are to choose the sequence that minimizes the quantity above.

The problem may be solved by straightforward enumeration when n and m are small. The efficient alternative is to define the dynamic programming recursion equation and use backward induction. To this end, let $a_i$ and $b_j$ be the first activity and toxicity screen, respectively, that remain to be passed through. Call this stage (i,j). Then define F(i,j) to be the expected cost of submitting a compound to the optimal sequencing of screens $(a_i, \ldots, a_m)$ and $(b_j, \ldots, b_n)$, given that it has already passed successfully through screens $(a_1, \ldots, a_{i-1})$ and $(b_1, \ldots, b_{j-1})$. Clearly,

$$F(m + 1, n + 1) = 0$$

$$F(i, n + 1) = C_{i,n+1}^a + P_{i,n+1}^a F(i + 1, n + 1)$$

$$F(m + 1, j) = C_{m+1,j}^b P_{m+1,j}^b F(m + 1, j + 1)$$

$$F(i,j) = \min\{C_{ij}^a + P_{ij}^a F(i + 1, j), C_{ij}^b + P_{ij}^b F(i, j + 1)\}$$

Solving these equations by backward induction, we obtain both the optimal policy and the corresponding expected cost $F(1,1)$. The procedure may easily be generalized to more than two series by adding additional subscripts.

A simple rule requiring no computation may be shown to hold if one makes the following assumptions. For all i, j, and k:

1. a. $C_{ij}^a = C_i^a$

   b. $C_{ij}^b = C_j^b$

2. a. $\dfrac{1 - P_{ij}^a}{C_i^a} \geqslant \dfrac{1 - P_{kj}^a}{C_k^a}$, $\quad k \geqslant i$

   b. $\dfrac{1 - P_{ij}^b}{C_j^b} \geqslant \dfrac{1 - P_{ik}^b}{C_k^b}$, $\quad k \geqslant j$

3. a. $(1 - P_{ij}^a) \leqslant (1 - P_{ij+1}^a)$

   b. $(1 - P_{ij}^b) \leqslant (1 - P_{ij+1}^b)$

All of these assumptions may well hold in a pharmaceutical setting: Assumption 1 means that the cost of passing a compound through a screen is independent of how the series a and b are interspersed in this screen. Assumption 2 means that the probability of failure per unit cost decreases as a compound passes through either the sequence of activity screens or the sequence of toxicity screens. Assumption 3 fails to hold only if there is a *positive* correlation between success at one of the activity screens and success at one or more of the toxicity screens. Thus, in particular, it will hold if the activity and toxicity series are independent of each other, or if all the correlations between the screens of one and the screens of the other are nonpositive.

Setting $v_{ij}^a = (1 - P_{ij}^a)/C_i^a$ and $v_{ij}^b = (1 - P_{ij}^b)/C_j^b$, one may show that under the assumptions above, it is optimal to select screen $a_i$ at stage (i,j) if and only if $v_{ij}^a \geqslant v_{ij}^b$. Thus at each stage we should select the screen that gives the highest failure rate per unit cost. For example, if the costs and success probabilities for the three activity screens are (1,10,100) and (0.10, 0.30, 0.50), respectively, while for the two toxicity screens they are $(5, 25)$ and $(0.05, 0.80)$, it is optimal to sequence these screens as $(a_1, b_1, a_2, b_2, a_3)$, yielding an expected cost of $3.75$ per compound. This may be compared to the sequence $(a_1, a_2, a_3, b_1, b_2)$, for which the expected cost per compound is $5.375$, approximately $52\%$ more.

## REFERENCES

Bergman, S. W. (1981), "Acceptance sampling: the buyer's problem," Ph.D. dissertaton, Yale University.

Bergman, S. W. (1985). "Allocation of resources to a sequence of screens," in preparation for publication.

Gittins, J. C. (1980), *Sequential Resource Allocation, A Progress Report*, Mathematical Institute, Oxford University.

Gittins, J. C. (1982), "Forwards Induction and Dynamic Allocation Indices," *Proceedings of NATO Conference on Deterministic and Stochastic Scheduling, Durham, 1981*, Reidel, Amsterdam.

# 4

## R&D PROJECT SELECTION METHODS

### 4.1. INTRODUCTION

One of the major responsibilities of managers of pharmaceutical research is the selection and funding of research projects. With research funds becoming increasingly scarce and more accountable, attention in the last 15 years has focused increasingly on finding aids that will improve the quality of these decisions (see, e.g., Dean, 1968, and Kadane and Simon, 1977).

The primary issues directly involved in exercising this responsibility are the selection of which projects to fund, and the determination of the levels at which they should be funded. With unlimited nontransferable resources, the solution to this problem would be to fund all proposed projects at the maximum level. But characteristically, there are more potential projects than there are resources available. The typical limiting factors include laboratory space and equipment, skilled personnel (chemists, biologists, and physicians, with various specialisms in molecular biology, molecular pharmacology, immunochemistry, cell biochemistry, etc.), and money. This is a more satisfactory situation than it may seem. As pointed out by Dean (1968), it is important that there exist more potential projects than can be accommodated. This will permit the manager to select those which will contribute most to the productivity of the firm. In fact, the absence of such a choice implies that more resources should be diverted to idea generation, a topic discussed further by Dean (1968).

Literally hundreds of papers have been published on the R&D resource allocation problem. Three journals, *IEEE Transactions on Engineering Management, R&D Management,* and *Research Management*, devote a large proportion of their pages to this topic. In addition several other journals, including *Management Science, Operations, Research,* and *Journal of the OR Society*, occasionally publish related articles. Major review papers have been published by Baker and Pound (1964), Baker and Freeland (1975), Cetron et al. (1967), and Souder (1972, 1973b, 1978). More specialized reviews have been published by Augood (1973), Gear et al. (1971), and Moore and Baker (1969b). A recent addition is a book on the management of research and innovation edited by Dean and Goldhar (1980). This includes a review article by Winkofsky et al. (1980).

In reading the literature, two principal facts emerge. The first is that very few of the models proposed have been tested in the field. The second is that even fewer have been adopted for regular use. This has caused considerable concern among those management scientists researching in the field of R&D project selection. It is, of course, important to understand some of the probable causes of this failure.

One factor is that the corporate environment often has not been favorable for the successful implementation of quantitative project selection models. Management style may be one important aspect of this factor. Faust (1971) points out that "a director having a prolonged exposure to, or philosophical affinity toward, the academic research environment may resist pressures from marketing or other functional areas, and also oppose the institution of various quantitative business analysis techniques in project planning and selection methodology. He may feel that the best research result will evolve from hiring the most qualified scientists and giving them maximum freedom to pursue research in their area of expertise." Faust found that such a management style characterized several major pharmaceutical firms.

Equally important in promoting the introduction of project selection techniques is the existence of an efficient communication network within the firm. Quantitative techniques require the integration of information from several sources that transgress the usual discipline boundaries. A research organization characterized by pockets of isolated units cannot contribute effectively to this task.

Corporate structure has a direct bearing on these factors. Typically, pharmaceutical laboratories are organized according to a matrix in which scientific personnel are grouped by disciplines (rows of matrix) and assigned to projects (columns of the matrix). That matrix may more closely resemble either a discipline-oriented coordination matrix in which the primary allegiance is to the discipline group, or a goal-oriented leadership matrix in which the primary allegiance is to the project. Stucki (1980), as Director of Pharmaceutical Research and Planning at Upjohn, found that the latter organizational structure increased communications between various disciplines, and fostered an attitude of the type that Faust described as being favorable for project selection models.

Other researchers (e.g., Souder, 1972a,b) have mentioned two other attitudinal elements which have prevented successful implementation of quantitative selection methods. one is the prior belief that these methods depend on very uncertain estimates, and that quantification simply compounds uncertainties with uncertainties, giving a mathematical result which suggests a precision that is not there. At the other extreme is the prior expectation that quantitative techniques will, in fact, resolve the allocation problem in an optimum fashion. The first attitude will prevent implementation, and the second will lead to disappointment when a quantitative procedure is implemented.

Our review of the corporate environment must be complemented by the acknowledgment that there are several deficiencies in the models themselves. Foremost among these is a lack of realism. This difficulty stems from the very nature of the selection problem itself. In an early review of resource allocation models, Pound (1964) stated that project selection was a relatively unstructured problem. In a more recent review, Moore and Baker (1969b) reiterated this theme. The difficulty is one of isolating the allocation problem from the numerous factors that influence and guide company policies in general. These include the need for company prestige, departmental loyalties, coalitions, and divergencies in viewpoint. A successful manager must take all these elements into consideration. Yet they are not easily modeled. Several authors (see, e.g., Souder and Souder) have noted that selection models have been especially weak in their organization behavioral content. Souder (1977) lists 10 differences commonly found between the real-world environment and the management science view. To overcome some of these disparities, he defines a behaviorally oriented procedure for project selection.

However, from another viewpoint it is questionable whether it is useful to try to model and incorporate such factors. In our view, project selection models are probably most useful when they

are normative models; that is, they describe what will be the results of a specific selection portfolio with respect to a given well-
defined set of criteria. If this is the way that they ought to be
used, this has certain implications about how they might be struc-
tured. In particular, they should be interactive and capable of
answering the "what if" type of question. This will permit the user
to explore the impact of certain alternative portfolios and balance
this against the nonquantitative considerations which are part of
the research environment. Several other management scientists
agree with this conclusion, among them Baker and Freeland (1975).

How successful have selection models been as normative
models? Again, to judge from their usage, not too successful.
Souder (1978) interviewed 26 R&D administrators (vice-presidents
and directors) and 26 R&D management scientists to ascertain
what performance criteria and characteristics were appropriate
for evaluating the suitability of (normative) R&D project selection
models. Judged most important was realism, followed by flexibility. Realism here means the ability of the model to model accurately the (nonbehavioral) elements of the real world. The most
elaborate of the mathematical programming models rated in Souder's article did not score above 50% on this criterion.

The conclusion to draw from the comments above is not that
quantitative project selection modeling has little to offer. We are
convinced otherwise for a very simple reason. Comparing projects
necessarily involves comparing some quantitative features characterizing these projects. These may include the probability of success, the expected time to completion, or the expected rate of return on successful completion. As comparisons of this type must
be made, it is better that they be made in an explicit and systematic fashion. Not only does this yield comparable figures for all
projects, but it also helps to pinpoint the type of information required for obtaining a realistic, if not necessarily precise, assessment of the relevant quantities affecting the promise of a project.
The need for a systematic procedure is especially evident in exploratory research, which involves a great deal of uncertainty.
It is well documented within the literature on personalistic probability estimates (see, e.g., Savage, 1970) that it is exceedingly
difficult for unaided intuition to combine in a rational fashion the
various sources of uncertainty associated with a project to arrive
at an overall assessment for the project as a whole. Established
modeling techniques may assist in this task.

In conclusion, the successful implementation of a quantitative
project selection technique appears to depend on two primary
factors. First, a realistic understanding by the manager of the

type of information that these models can provide and how this information can be integrated into the overall assessment process. Second, that the models themselves successfully model the essential features of the research and development process, and thus provide information that the manager will find useful.

In the pages that follow this introduction a rather detailed presentation is given of many of the project selection models found in the literature. Not all of them are applicable to pharmaceutical research. However, they are included here for two reasons. First, as the project manager becomes increasingly familiar with quantitative models, he or she will encounter references to some of these models and will wonder what they are about and if they are relevant to his or her needs. Second, even though not directly applicable, some of these models contain useful features that the manager may wish to incorporate into a model of his or her own. Here, in the remainder of this introduction, a nontechnical overview is given of the models found in the main body of this chapter, with respect to both their applicability and their ease of use.

The first model discussed in Section 4.2 is the checklist. This is simply a listing of all the factors that are believed to be relevant to assessing the worth of any project, together with a weighting scheme designating the importance of these factors. Once the items appearing on the list are established, each project is assigned a score with respect to each of these items. Although quite simple in format, the checklist is a useful selection tool with wide applicability. Its virtues are: (1) it explicitly identifies all factors relevant to the project selection task; (2) it explicitly weights the importance of each of these factors; (3) each project is assessed with respect to all of these factors, pinpointing strengths and weaknesses and where there is a lack of information; (4) it is easy to use; (5) it is inexpensive; and (6) it is easy to understand.

The main task associated with implementing a checklist is determining the relevant factors and their relative weights. These factors may be either quantitative (e.g., probability of success) or nonquantitative (e.g., prestige value). This task is of fundamental importance in its own right, and should be done prior to implementing any project selection model. The proper continued use of the checklist requires a flow of information from various departments, permitting scores to be assigned to each project on each criterion. The estimates are not necessarily precise, but are the best currently available. The usefulness of the checklist may be increased by including a measure of the precision with which the estimates are made.

A derivative of the checklist is the profile chart. It displays graphically the information contained in the checklist and is one of the most useful semiquantitative project selection techniques. Figure 28 gives an example of the range of information a profile chart can display and the ease with which it permits the user to assimilate this information. There, the height of the box for each factor corresponds to the relative weight assigned to it; the left end, the middle break, and the right end of the box define the 25th, 50th, and 75th percentile of the distribution associated with the score's estimate; and the tone of gray highlights the uncertainty of the estimate. Hence, in this example, we note that the figure for projected sales is not known with any great precision but is thought to be high, while the scientific manpower requirement is thought to be very likely near the average requirement of previous projects.

From either the checklist or the profile chart a score index may be derived which will summarize the overall worth of the project. It is typically obtained by summing over the weighted scores of the project. Being a single measure, this permits the user to unambiguously rank projects in terms of their apparent overall attractiveness. However, for a detailed understanding of the strengths and weaknesses of a project, the profile chart is superior. It follows that scoring indices are most useful when such detailed knowledge is not necessary. For example, they can be used effectively to eliminate noncontenders by removing from consideration all projects that score below some threshold.

Much more useful than scoring indices are profitability indices, discussed in Section 4.4. Using standard cost accounting and decision-theoretic concepts they combine various quantitative elements in a logical way to obtain a measure of the project's financial worth. Many such measures have been discussed in the literature. These include indices proposed by Olsen, Pacifico, Disman, and Dean-Nishry (see, e.g., Augood, 1973). The latter is interesting for it takes into account the effect that the introduction of a new product would have on existing company products. However, none of these adequately model an essential feature of exploratory research—the possibility of abandoning a project at any stage of the research process, thus not only avoiding future losses but also freeing resources for profitable use elsewhere. Pearson (1972) discusses the importance of this multistage feature of research for the selection problem, and points out that it is economically desirable "to allocate some funds to projects which have only a low expectation based on current estimates of their probability of research success if these estimates can be proved to be true or false at little cost." Using a decision tree analysis, he suggests a relatively simple profile index that merits consideration. The information required for its use are estimates of the cost and the probability of successful completion of each research phase, and an estimate of the return if the entire project is successful. The modified version given in this report also takes into consideration the effect of preemption by a competitor, but then the likelihood of this eventuality occurring at any research stage must also be estimated.

A similar index, called the dynamic allocation index (DAI), has been proposed by Gittins (1979). This has been shown to be optimal for a wide range of problems involving the sequential allocation of resources arising in statistics and scheduling. It is characterized as the maximal rate of expected net (discounted) return per time unit that can be achieved by using a stopping rule. In the form described in this book it is not computationally any more difficult to obtain than the Pearson index, and it requires essentially the same type of input estimates.

Although useful project selection tools, it is important to note that profitability indices do not allow for benefits that are of a nonfinancial nature. Hence it is worthwhile to complement them with a nonfinancial index. This may be done by splitting the checklist into two parts, one concerning the profitability of the project, and the other detailing nonfinancial benefits. A scoring scheme may be adopted for the latter, while a profitability index of the type recommended is used for the former.

The techniques described above aim mainly at defining the relative merits of individual projects. The fact that ranking procedures based on individual merit measures do not lead to optimal portfolios under resource constraint has been pointed out by several authors (see Section 4.5). The difficulty may be understood in terms of packing a knapsack with items of different value and volume, with the objective of maximizing the total value of the items packed. The profitability indices would call for selecting in descending order those items with the greatest value per unit volume. However, this may lead to "end effects" when none of the remaining items can be squeezed into the knapsack, thus leaving unused space. In terms of project selection, the unused space may represent underutilization of senior scientists, technicians, laboratory facilities, or research funds. To overcome this specific facet of portfolio selection with constraints, a number of integer and linear programming models have been proposed. With the programming packages which are commercially available, they are not difficult to use. The type of input parameters required are an itemization of the resource constraints over the various budget periods that encompass the lifetimes of the projects in question, and an estimate of the return upon completion. One of the more interesting outputs, other than the optimal portfolio satisfying the constraints, is the marginal value index computed for each resource constraint. These give a measure of the increase that may be obtained in the objective function (return) by a small increase in the respective resource constraints.

However, there are a number of drawbacks associated with many of these models that cast doubt on their applicability to an exploratory research environment. Foremost is that the possibility of unyielding resource constraints appears to be a secondary feature of the resource allocation problem compared to the following essential features which are often not modeled: the multistage character of the research process, the nonlinearity of the return function, and the myriad of uncertainties associated with costs, likelihood of success, time to completion, preemption, and benefits upon completion.

Several attempts have been made to capture some of these features within a linear or integer programming framework, although not all within one model. Stochastic programming (see Section 4.6) models the multistage character of the research process, but not the associated uncertainties in costs. The main input requirements for it are a specification of the costs, success probabilities, and resource constraints associated with each research stage, and an estimate of the benefits upon successful completion. As an alternative we suggest a simpler model which permits uncertainty costs. The input requirements are the same. Other examples are given in Section 4.7. Allen and Johnson (1970) use simulation to trace out the effects on the optimal policy of uncertainties regarding benefits. This methodological approach is attractive, for not only does it give a collection of nearly optimal policies to choose from, it also provides a sensitivity analysis with respect to the input data that are being varied. One drawback is that it is not easily programmed, and some familiarity with simulation is required of the person in charge of implementing the model. Finally, chance constrained programming (see, e.g., Charnes and Cooper, 1963) permits uncertainty in costs, but does not, for example, capture the sequential aspect of the research process. We show that by a transformation this model may also be solved as an integer programming problem, and hence is not more difficult to implement than, say, the stochastic programming formulation discussed previously.

In principle, a much more flexible technique for adequately modeling the project portfolio problem is by the use of dynamic programming. This general method can model the multistage character of research, and can accommodate a variety of uncertainties, as well as nonlinear objective functions. However, three difficulties occur in practice. First, unlike linear and integer programming, there are no general portmanteau algorithms that can be used to solve the dynamic programming recursion equations. Any dynamic programming model must thus be individually programmed. Second, to obtain computationally feasible models it is often necessary to forgo many of the modeling refinements that are theoretically possible. Third, considerable sophistication is required of the user both in understanding the model and in interpreting the result.

In Section 4.8 two dynamic programming models are discussed. In the Hess model (1962) it is assumed that research and development of a selected project is continued until one of the following two milestones is reached: the project achieves "technical success," or a predetermined time horizon is reached without success, at which time the project is terminated as unsuccessful. This may not be unreasonable. Ideally, the reward should be a decreasing function of the time of completion. The Hess model allows for this. Also, the probability of success should be a function of the amount of resources devoted to the project. This is more difficult to model, and Hess offers two simplifying formulations. One is that lack of success in the preceding period implies that one is "starting from scratch" in the present period. Hess speculates that this assumption may not be severely violated in pharmaceutical research. The other assumes that the probability of success in the present period is an exponentially weighted sum of past research expenditures. With these assumptions it is possible to use backward induction to obtain the optimal allocation over time for a single project without budgetary constraints. This may very well be a useful piece of information. However, the primary emphasis is on the impact of a budgetary constraint upon the allocation problem. Assuming such a constraint for the first time period only, Hess derives the optimal allocation between different projects. Conditional on this result, it is then possible to derive the corresponding allocation over time within each project. The input parameters required by the Hess model are more extensive than the linear and integer programming formulations considered previously. In particular, both an estimate of the reward as a function of time of completion, and a coefficient defining the probability of success for the present period as a function of past expenditures, are required.

Rosen and Souder (1965) offer a modification of the Hess model. First, the total expected discounted gross profit is simplified by assuming that it is not a function of time. Although this reduces the need to estimate parameters, it is a rather doubtful assumption. Second, the probability of success by time t is assumed not to depend on past expenditures. This is similar to the first formulation given by Hess. Third, constraints are put on the maximal and minimal amount to be spent on a project during each successive planning period. This is useful and leads to no loss of generality. Fourth, the form of the probability of success as a function of current expenditure is allowed to be arbitrary. This is an improvement, but requires more input parameters to be specified.

An alternative to the foregoing two models is offered by Atkinson and Bobis (1969). While formulated as a nonlinear rather than a dynamic programming problem, it is quite sophisticated and offers many attractive features. Like Hess, and Rosen and Souder, Atkinson and Bobis consider the problem of achieving an optimal allocation of resources within a project with respect to a sequence of budgets encompassing its life, as well as that of achieving an optimal allocation of resources between various competing projects. They see the former as one of balancing two conflicting costs. First, delaying the completion of a project will severely influence its market value when it is completed. Second, rushing a project will necessarily lead to inefficient use of research resources. These two features, together with the assumption that the probability of completion at time t is a function of past "effective" allocations, give their model a realism that is not possessed by many others. However, the model is very demanding on the user. Among the many parameters that must be estimated is a sequence of efficient allocations over the life of each project that define the maximum amount by which expenditures can be increased in each period without experiencing decreasing marginal returns on this increased investment.

Although at times very sophisticated, almost all of the above mathematical programming formulations share a common methodological perspective which seriously affects their usefulness. Their objective is to discover an optimal allocation with respect to a certain set of criteria, and consequently their primary, and often only output is the optimal allocation. This type of information is very difficult to integrate into the overall project selection process. If the derived optimal allocation is not possible because it conflicts with other nonfinancial objectives, the research manager has obtained very little useful information. Ideally, what he or she would wish to do is to explore the financial implications of alternative portfolios which are feasible, and to choose among them. Typically, these programs do not permit this type of interactive exploration.

A new approach has been adopted by Gittins and Roberts (1981) which overcomes this defect while simultaneously permitting models of unusually high levels of realism to be formulated. The principle is not one of determining the optimal allocation, but rather one of determining the financial implications of a given portfolio, together with indicators of how this portfolio may be improved. If one wished, one could in fact use these indicators to find a sequence of alternative portfolios that would converge to an optimal allocation. However, the character of the Gittins and Roberts model is more that of a sophisticated profitability index. For example, for a given portfolio the following financial and performance parameters are computed:

1.  Expected total present value of cash flows resulting from the selected projects, excluding research costs, and the contribution each project makes to this total
2.  The expected total discounted effort that is required by these projects, together with the contribution of each project to this total
3.  The expected present value generated per unit of discounted effort allocated to existing projects

In addition, for each project and each budgetary planning period, the following performance measures are given:

1.  The probability of completion by that period
2.  The probability of preemption by that period
3.  The probability that work is still in progress at the end of that period

These parameters may be used to assess the risk associated with the allocation plan, and the probable availability of resources for other projects at each time period in the future.

The indicators that allow the user systematically to improve his or her portfolio are the following:

1.  A marginal profitability index for each project and period
2.  A marginal profitability index for the portfolio as a whole for each time period

Some of the assumptions built into the model are not very different in character (as opposed to detail) from those used by Atkinson and Bobis. First, the value of a project is a function of its time to completion. Second, the probability of success by time t is a function of "effective effort" by time t. Some additional features are parameters for the rate of obsolescence and preemption. More unusual is the fact that the cost of resources is computed internally in terms of the cash that could be generated if they were applied to other projects. This frees management from having to assign an explicit value to the resources being expended; this may be an advantage since it is often difficult to determine a monetary value.

The Gittins-Roberts model is available as a FORTRAN-based package called RESPRO. The primary inputs required for each project are:

1. The long-run rate of return on capital employed by the firm
2. The time required for a newly formed research team to achieve full efficiency
3. The ultimate discovery rate per unit effort (i.e., the reciprocal of the expected time to find a potential new product when full efficiency is reached and the rate of effort is one unit per time unit)
4. A set of parameters defining the relationship between the rate of effort per time unit and the rate of progress per time unit
5. The new product's value assuming that it were discovered now
6. Obsolescence rate
7. Preemption rate

Given that the manager can supply these parameters, RESPRO should be no more difficult to use than other package-based programs. It is currently being used on an experimental basis by one company and is described in Section 4.9.

Other approaches to the problem of finding portfolios that satisfy more than one objective also exist. For example, management may be interested in achieving both a high profit and a high sales volume. These are the multiobjective models, some of which are discussed in the last section of this chapter. However, their degree of sophistication for handling uncertainty seem inadequate for a pharmaceutical research environment. These methods are discussed in Section 4.10.

## 4.2. CHECKLISTS AND PROFILE CHARTS

A first step in establishing a selection model is that of determining which elements contribute to the ultimate success or failure of a project. This probably could not and should not be done by one person. Rather, each relevant decision-maker should participate in this process, specifying all criteria that he or she believes to be relevant to the evaluation and selection of projects for research.

Moore and Baker (1969a) give a good introductory tutorial in the art of constructing a checklist. They suggest that the composite list obtained should have the following properties:

1. It should be *complete*, to guarantee that no important evaluation factors are overlooked in the analysis.
2. Every item on the list should be truly relevant.
3. Every item on the list should be measurable in the sense that a method and a scale for obtaining a measure of project performance with respect to this item either exists or can be constructed. The particular measure may be "natural" or "artificial."

To achieve these three properties a number of meetings are required of all those concerned, to resolve conflicting viewpoints about what is relevant and what is not.

In initially setting down the criteria that are to appear on the checklist, it is helpful to be familiar with factors other authors have found relevant. Moore and Baker, in considering a general research environment, suggest the following items: probability of success, project costs, income or cost savings, the timing of income and cost streams, level of technical and managerial familiarity with the research area, budget levels, market penetration, time to completion, and strategic need. Faust (1971) proposes the following as important factors in a pharmaceutical environment:*

*Scientific factors*

1. Interrelationships with other research activities—synergistic advantages, or competition with other programs
2. Probability of achieving project objectives
3. Time required to achieve project objectives
4. Impact on balance of short- and long-term programs within research
5. Estimated cost of the project in the coming year and to completion

---

*Reprinted with permission of Research Management, Technomic Publishers, Inc., Lancaster, Pennsylvania.

6.  Utilization of existing research talent and resources
7.  Value as a means of generating experience and gaining technical expertise in a field: a foundation for future research activities
8.  Need for critical mass of expertise and activity to ensure progress
9.  Elasticity of resource input and probable output relationships
10. Patentability or exclusivity of discoveries from project
11. Competitive research effort in the area—in academic and government research centers

*Marketing considerations*

1.  Projected sales and profits from effort
2.  Relationship to need as reflected by current state of consumer satisfaction
3.  Status and efficacy of current competitive products or means of meeting consumer need
4.  Compatibility with current marketing capabilities and strengths
5.  Influence of new competitive products under development

*Organization and other elements*

1.  Relationship to activities at other research centers or units within the company
2.  Timing of project with respect to other activities in marketing, research, etc.
3.  Manufacturing capabilities and needs
4.  Prestige and image value to the company
5.  Effect on organization esprit de corps and attitudes
6.  Impact of governmental, public opinion, and other environmental pressures
7.  Alternative uses of scientific personnel and facilities if project dropped after a few years
8.  Moral compulsion to develop drugs meeting medical need but having low or no profit potential

Not all of the factors above are necessarily relevant, and factors other than those above may be important. In addition, we suggest that the following items be incorporated in the list:

1.  The number and expected timing of the milestones of the project if it proceeds to completion
2.  The expected costs and probabilities of reaching each milestone

The resulting composite list of criteria will, of course, be unique to the research environment to which it corresponds.

The next step in producing the checklist is one of defining the scales of measurement for each criterion on the list. It may be the case that there exist several different "natural" measures for one criterion. Any of these may be chosen, but as a source of input into other selection models it is convenient to express quantities in monetary units if that is one of the possible natural measures.

The problem of how to choose the checklist's units of measurement has been considered by Moore and Baker (1969b). Their method, the environmental model, is quite sensible and overcomes some of the apparent arbitrariness exhibited by other schemes. The method may be described as follows. For each criterion one determines a probability distribution for the set of values that previous projects within the firm have had for this criterion. It is assumed that the resulting curve is a bell-shaped curve resembling the normal distribution. The mean $\mu$ and variance $\sigma^2$ are determined for this distribution, and the original measure, x, for the criterion is transformed into the new measure $y = \pm(x - \mu)/\sigma$, where the sign is chosen such that positive values correspond to favorable, and negative to unfavorable, scores.

The process is illustrated in Figure 22. The measure y may be made discrete by appropriate subdivisions of the continuum it spans. More and Baker have found that the nine different scores shown in Figure 22 are sufficient in terms of both effective range and power of discrimination.

The structure of the checklist is now complete. Each project is rated subjectively on each of the criteria on the list. Comparisons are made between projects by comparing their scores on the various attributes contained in the checklist. The inherent advantage of this simple scheme is that it effectively summarizes all relevant information about each project, making a more systematic comparison possible between different contending projects, and even between a present project and the average performance of past projects.

Rather than keeping the information in tabular form, it may be more effectively displayed graphically. Harris (1961) has promoted this idea at Manosator. Such a resulting profile chart is shown in Figure 23.

A serious disadvantage of this construction, especially when displayed in profile form, is the natural tendency of the user to correlate subjectively the number of high scores a project receives with its worth. Not all categories are in fact of equal im-

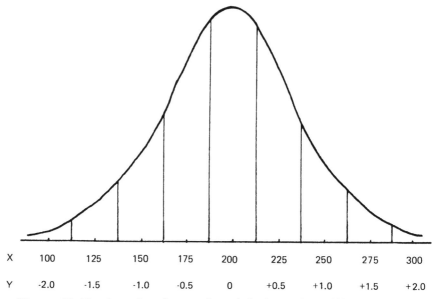

| X | 100 | 125 | 150 | 175 | 200 | 225 | 250 | 275 | 300 |
|---|-----|-----|-----|-----|-----|-----|-----|-----|-----|
| Y | -2.0 | -1.5 | -1.0 | -0.5 | 0 | +0.5 | +1.0 | +1.5 | +2.0 |

**Figure 22** Total project income in original x values ('000 Dollars) and transformed y values. (Reprinted with permission of the Institute of Management Sciences, © 1969.)

portance. For example, in Figure 24 two profile charts are presented of the same project. In the second of these charts the first two categories are each broken down into two identical categories, creating six altogether. Clearly, the second profile looks more promising.

In recognizing this problem, Moore and Baker recommend that criterion overlap should be held to a minimum. However, it is clear that what is really required is that each criterion be assigned a weight, and that the profile chart display this weight. Several methods of weighting criteria for individuals or groups are available. These range from simple rank-ordering schemes (see Section 4.3 for an example of their construction and usage given by Dean and Nishry) to partial or complete paired comparisons (Buel, 1960; Churchman et al., 1957). Moore and Baker point out that in two independent testing situations it has been determined that a selected set of these methods produce similar weights, but that the simple ranking method is by far the easiest to use.

The corresponding weight may be listed adjacent to each criterion on the checklist. If a profile chart is used, it should be

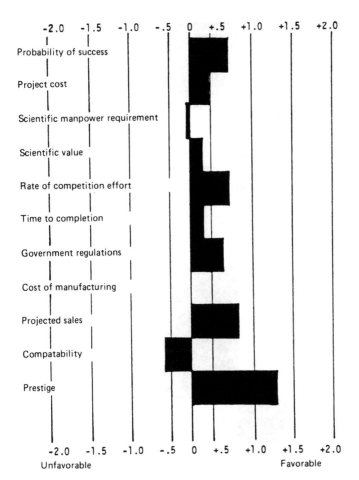

**Figure 23** A profile chart.

modified by having the width of the bar corresponding to a criterion proportional to the weight attached to that criterion. With the first chart in Figure 23 appropriately modified to reflect the weighting 5, 5, 6, 3, 3, 3, 5, 3, 8, 4, 1, we obtain the profile chart shown in Figure 25.

The profile chart may be made to display additional refinements. Rather than obtaining a single-point estimate of a project's rating with respect to a criterion, one may wish to obtain an estimate of its probability distribution with respect to this criterion. Such a distribution can often be effectively summarized by

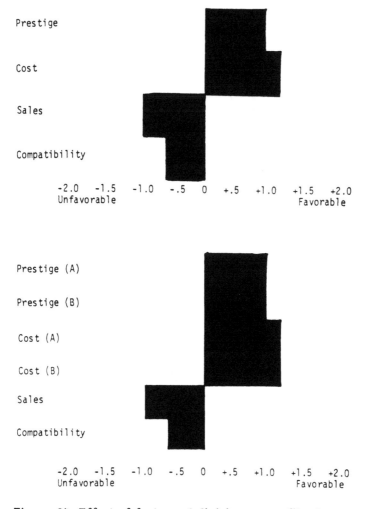

**Figure 24** Effect of factor subdivision on profile chart.

the median and the first and third quartiles. In a different context Tukey (1977) uses these three statistics to display distributions graphically in the form of "box plots" (Figure 26). The

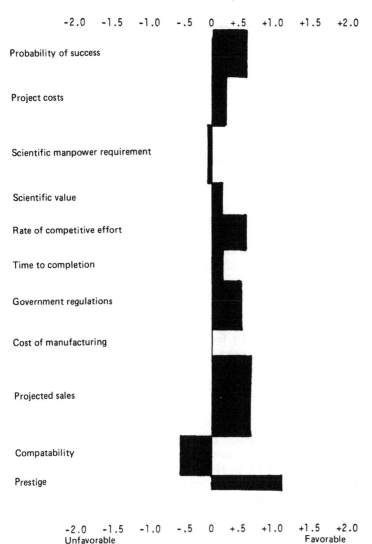

Figure 25  A relevance weighted profile chart.

chart in Figure 27 illustrates this use of box plots.  One drawback that is immediately apparent is that criteria with high variance are given greater visual impact.  To correct for this visual effect, the final chart in Figure 28 (See also Bergman, 1981) uses box

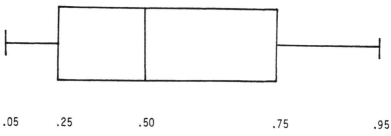

.05      .25          .50                    .75               .95

**Figure 26** Tukey box plot with percentiles of distribution displayed.

plots together with an additional refinement: various tones of gray are used to display the credibility, which is defined to be inversely proportional to the range between the quartiles. Consequently, no additional information is actually contained in Figure 28 beyond that contained in Figure 27. However, it helps the eye to assimilate the information that is displayed in the correct manner.

With the refinements suggested above, a profile chart can be made to reveal almost all known factors about a project, pinpointing both the strengths and weaknesses, and where additional information should be collected.

## 4.3. SCORING INDICES

A score index is a single number derived from a checklist to describe the overall worth of the project. Being a single measure, this permits the user to rank projects unambiguously in terms of their overall attractiveness. The standard procedure in selecting a portfolio is then to choose projects by selecting them in the order of their ranking until insufficient resources are left for any of the remaining projects. An introduction to this topic may be found in Augood (1973).

The various scoring methods suggested in the literature may be classified into two groups: those for which the function defining the transformation is additive, and those for which it is multiplicative. Moore and Baker (1969b) argue that the former shows greater consistency with other project selection methods and is less sensitive to errors in estimation. As a result, their suggested procedure is simply one of multiplying the point estimates for each criterion by the weight of that criterion and then

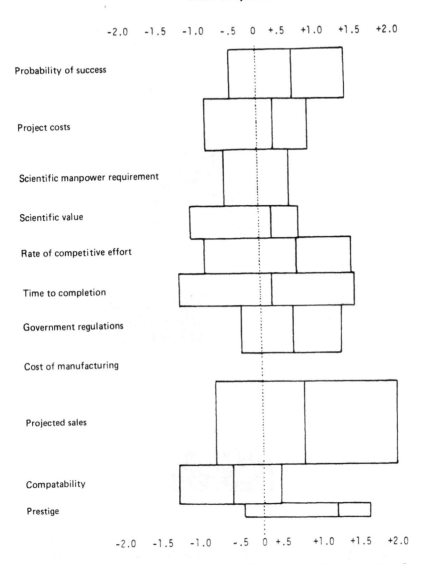

**Figure 27** A relevance-weighted profile chart with box plots of score estimates.

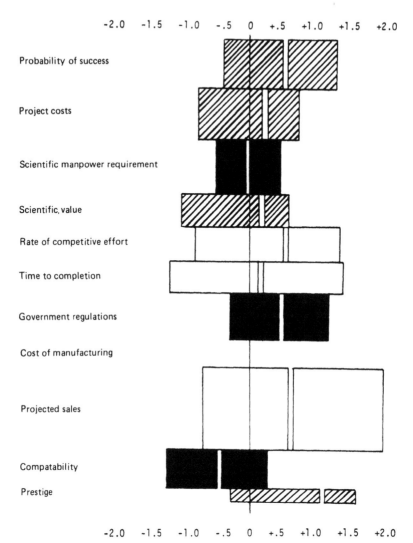

**Figure 28** A relevance-weighted profile chart with shaded box plots of score estimates.

summing. This corresponds to determining the area of the shaded region above the zero line in Figure 24 and subtracting from it the area of the shaded region below the zero line. Since the scores on each criterion in the checklist are normalized to have an average value of zero, projects whose overall rating is greater than zero are better than average, and conversely those with an overall rating less than zero are worse than average. In addition, one may determine how much better or worse than average a project is by directly constructing the overall score index distribution. Under the assumptions of normality for the distribution of each criterion and weights $w_i$ for criterion i (i = 1,...n), this distribution will be normal with mean 0 and variance $\Sigma_{i=1}^n w_i^2$. In view of this, it is convenient to rescale the score by dividing it by $\Sigma_{i=1}^n w_i^2$, to obtain an overall score which has the standard normal distribution.

Other scoring models discussed in the literature are those by Cooper (1978), Dean and Nishry (1965), Gloskey (1960), Hart (1967), and Paolini and Glaser (1977). As the Dean-Nishry model was rated favorably by Souder (1973) in his review of selection models, we discuss it in more detail below.

The Dean-Nishry index is an additive scoring model consisting of two parts: a technical score I' and a market score I''. These two parts are initially considered separately, and for each a checklist is created, consisting of relevant factors {i'}, {i''} and corresponding weights {$w_i'$}, {$w_i''$}. Dean and Nishry describe the determination of the weights as follows: "The review board members, acting independently, rank order the technical and market factors. The rank orders are converted into numerical values, assuming equal intervals between adjacent ranks. These values are then averaged across the review board members, assuming approximately equal degrees of knowledge." In terms of the normalization we suggested for the Moore-Baker model, the values above should be scaled so that their squares sum to 1. The technical and marketing scores for the jth project are then defined to be

$$I_j' = \sum_i w_i' y_{ji}' \quad \text{and} \quad I_j'' = \sum_i w_i'' y_{ji}''$$

where

$w_i'$ = weight for the ith technical factor

$y_{ji}'$ = value for technical factor i in the jth project

$w_i''$ = weight for the ith marketing factor

$y_{ji}''$ = value for marketing factor i in the jth project

An overall score $I_j$ for the jth project is defined to be

$$I_j = aI_j' + bI_j''$$

where a and b are weighting factors such that $a^2 + b^2 = 1$. Presumably, the technique for deriving values for a and b are the same as those for deriving $\{w_i'\}$, $\{w_i''\}$. Formally, the Dean-Nishry model does not differ from the Moore-Baker model.

Some general comments regarding scoring models are appropriate at this time. First, they are a useful way of summarizing some of the information contained on a checklist. However, for a detailed understanding of the strengths and weaknesses of a project proposal, the profile chart is superior. It follows that scoring indices are most useful when such detailed knowledge is not necessary. For example, they can be used effectively to eliminate noncontenders by removing from consideration all projects whose score is lower than some threshold. What this threshold value should be will depend on the proportion p of all proposals that are ultimately accepted. If one felt that twice as many should be passed on for further consideration as the number that ultimately will be accepted, then on the average any project with a normalized score less than $I_0 = \Phi^{-1}(1 - 2p)$ should be eliminated from contention, where $\Phi^{-1}(\cdot)$ is the inverse of the standard normal distribution function.

Second, the scoring method can be made more informative by extending it to include more than one summary statistic. For example, the scoring pair $(I_1, I_2)$ may be useful in identifying projects about which there is already sufficient information to make a decision. Using Figure 28, define $I_1$ as the sum of the expected values on each of the criteria by the formula

$$I_1 = \frac{\Sigma_i w_i m_i}{\Sigma_i w_i^2}$$

where $m_i$ is the second quartile (mean) for the ith criterion. Then define

$$I_2 = \frac{\Sigma_i [w_i (q_{3i} - q_{1i})]}{1.3498 \Sigma_i w_i^2}$$

where $q_{i1}$ and $q_{i3}$ are the first and third quartiles displayed in the chart. Not plot $(I_1, I_2)$ for all projects on graph paper. An example of such a plot is shown in Figure 29. With the dashed decision boundaries drawn as they are in this graph, projects 1, 3, and 11 should be rejected; projects 2, 4, 9, and 10 should be considered for funding; and more information should be gathered for the remaining projects.

## 4.4. PROFITABILITY INDICES

The strength of the scoring index is that it permits the simultaneous weighting of rather different qualitative and quantitative factors. This is also its weakness, for it implies that no sound theoretical relationship has been used to establish the appropriate

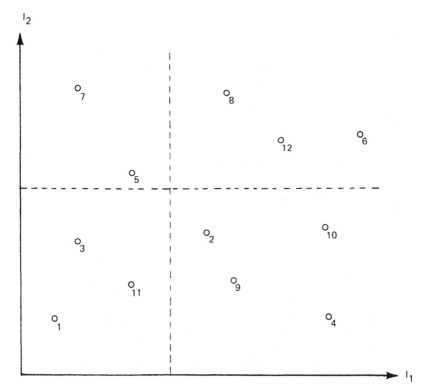

Figure 29  A mean and standard deviation plot for the scores of 12 projects.

functional relationship between the components of the index. The profitability indices overcome this weakness by sacrificing the former strength. They are restricted to analyzing those factors that can be weighted solely in terms of their contribution to the financial gains that may be expected from the project. With this restriction it becomes possible to apply cost accounting and decision-theoretic concepts to construct significantly more sophisticated measures of a project's (financial) worth.

Several different profitability indexes have been suggested in the literature. Among these are Ansoff (1964), Dean and Nishry (1965), Disman (1962), Gloskey (1960), Hart (1967), and indices due to Pacifico, and Teal, all of which are discussed in Augood (1975). Review articles discussing these and other indexes have been published by Allen and November (1969), Augood (1973), Baker and Pound (1964), and Souder (1973, 1978).

The two basic components of any measure of profitability of a project i are the expected reward $R_i$ and cost $C_i$ that will result from initiating this project. A measure of the net benefit from project i is therefore $B_i = R_i - C_i$. However, this quantity is not appropriate for ranking projects for the purpose of selection. The selection problem may be stated as one of selecting a subset $(i_1, \ldots, i_m)$ of the set of proposed projects $(1, 2, \ldots, n)$ so as to maximize the total net benefit

$$\sum_{j=1}^{m} B_{i_j}$$

subject to the budgetary constraint

$$\sum_{j=1}^{m} C_{i_j} \leqslant \text{total budget}$$

It is easy to show that the correct solution to this problem (apart from possible end effects, which are discussed in Section 4.5) is to rank the projects in decreasing order of the benefit/cost ratio $B_i/C_i$, and to select from the top of the list until the budget constraint prevents further selection.

Almost all profitability indices studied in the literature are variations of the benefit/cost ratio. They differ largely in how benefits and costs are defined.

Two of the simplest indices are those by Olsen and Pacifico. They are

$$I_{\text{Olsen}} = \frac{P_R \cdot P_D \cdot P_M \cdot S \cdot M \cdot N}{C_R}$$

$$I_{\text{Pacifico}} = \frac{P_R \cdot P_D \cdot P_M \cdot S \cdot M \cdot N}{C_R + C_D + C_M}$$

where

$P_R$ = probability of research success

$P_D$ = probability of development success

$P_M$ = probability of marketing success

S = annual sales volume if successful

M = profit per unit sold

N = product's life span (years)

$C_R$ = cost of research

$C_D$ = cost of development

$C_M$ = cost of marketing

The basic formula in both is

$$\frac{P_s \cdot B}{C}$$

where

$P_s$ = probability of success = $P_R \cdot P_D \cdot P_M$

B = benefit if successful = $S \cdot M \cdot N$

C = cost

The difference in C for Olsen (C = $C_R$) and Pacifico (C = $C_R$ + $C_D$ + $C_M$) may be attributed to two different types of budgetary constraint. For Olsen it is the research budget, whereas for Pacifico it is the research, development, and marketing budgets combined. If the overall objective is to maximize the expected profit to the firm, Pacifico's denominator is the appropriate one. However, this requires that the research, development, and marketing budgets be combined and treated in practice as one overall budgetary constraint. When this is not the practice, the denominator should be chosen to be the costs subsumed under the budgetary constraint limiting the projects that may be selected at this phase. In a research laboratory environment this is usually $C_R$ or $C_R + C_D$. Hence Olsen's index is the more appropriate of the two.

Disman modifies Olsen's formula to take into consideration the fact that future money is not equivalent in value to present money. Appropriately discounting income streams, he obtains

$$I_{Disman} = \frac{P_R \cdot P_D \cdot P_M [\Sigma_{i=1}^{N} R_i/(1 + r)^i]}{C_R (\text{discounted})}$$

where

$R_i$ = estimated revenue in the ith year

$r$ = interest rate

While highly rated in Souder's (1973) review of selection models, we should also note one unusual feature in the formula above. The index is not measuring expected net benefits, but rather, expected revenues. If our objective is the former, the appropriate modification would yield

$$I_1 = \frac{P_R P_D P_M [\Sigma_{i=1}^{N} R_i/(1 + r)^i] - C_R - C_D - C_M}{C_R}$$

where all costs are time discounted. The same observation may apply to Olsen's and Pacifico's indices, depending on whether or not the total benefit $S \cdot M \cdot N$ is defined to include research, developing, and marketing costs. Clearly, care should be taken in defining what is to be maximized, that is, what is to be meant by the total benefit.

Dean and Nishry (1965) develop a more elaborate index which takes account of the effect that the introduction of a new product would have on existing company products. Three different cases are considered:

1. A project, leading to a new product line in the company, that does not have any significant effect on existing profitability
2. A project, leading to a new product which is within an existing product line, that may have an effect on existing product profitability

3. A project, leading to a modification of an existing product, which replaces the unmodified product

The expected profitability of initiating the project is then defined to be

$$\tilde{I} = \begin{cases} I' & \text{for projects of type 1} \\ I' + \sum_j I''_j & \text{for projects of type 2} \\ I' + I''_j & \text{for projects of type 3} \end{cases}$$

where

$$I' = \left\{ \sum_{n=1}^{L} S(n)[M_S - C_{MF} - C_S]d^n - C_T \right\} P_S - C_D$$

$$I''_j = \left\{ \sum_{n=1}^{L_j} S^j(n)[M_S^j - C_{MF}^j - C_S^j]d^n \right\} P_S$$

and

$L$ = estimated sales life of new product

$S(n)$ = estimated number of units sold of new product in nth year

$M_S$ = unit sales price of new product

$C_{MF}$ = unit manufacturing cost of new product

$d$ = discount factor = $1/(1 + r)$

$r$ = annual rate of return on investment

$C_T$ = tooling costs for new product

$P_S$ = probability of success in creating new product

$C_D$ = development costs of new product

Except for $S^j(n)$, adding the superscript j to the quantities above means that they refer to existing product j. $S^j(n)$ is defined to be the marginal change in the sales of existing product j in year n due to the successful introduction of the new product. It may be negative in sign.

   The Dean-Nishry index for project selection is now obtained by forming the benefit/cost ratio

$$I_{\text{Dean-Nishry}} = \frac{\tilde{I}}{C_D}$$

The indices above, and others in the literature, may be fault-
ed for not taking into account two important factors that affect
profitability in a research environment. The first is that research
funding is a sequential process. Pearson (1972) discusses the
implication of this for selection, and illustrates his arguments with
the following example. Suppose that there are two projects, with
respective probabilities of research, development, and marketing
success of 0.1, 1, 1 and 1, 1, 0.1. Assume that their costs and
rewards are specified such that their benefit/cost ratios, as defined by one of the procedures above, are identical. It will not
follow, however, that these two projects are equally attractive.
This is because:

1.  In the first case there is only a 10% chance that expendi-
    ture subsequent to the research stage will be required,
    while the commercial benefits are certain if the research
    is successful.
2.  In the second case all the expenditure must be incurred
    to obtain a 10% change of achieving the estimated bene-
    fits.

Thus, although it may be unattractive to commit all the required resources to the second project, it may be quite attractive
to pay for the initial research required in the first project to determine whether or not the project will be commercially successful.
For example, most people would be unwilling to gamble $1 million
to win $2 million with a 50% chance. But almost all would be will-
ing to invest $1 to discover if one would win such a gamble, when
it is understood that one subsequently could choose to play if the
results were favorable and not play if they were not.

The sequential nature of resource allocation is thus an important aspect of pharmaceutical research. It permits a project to be
abandoned, not only avoiding future losses but also freeing resources for profitable use elsewhere. In the following we consider two different project indices incorporating this essential feature of exploratory research.

The first is obtained by modeling the sequential aspect of the
research process by the use of decision trees. In particular, assume that there are three different phases in the research and
development of a drug, and that these are expected to terminate
at times $t_1$, $t_2$, and $t_3$. Let the costs of these three phases be
denoted by $C_1$, $C_2$, and $C_3$, and let the respective (conditional)

probabilities of success at each of these phases be $P_{S1}$, $P_{S2}$, and $P_{S3}$. Successful development of a drug will presumably lead to marketing of the drug at, say, time $t_4$, and this will yield an expected reward of B. The computation of B may include several factors, such as the probability of commercial success, $P_M$; expected marketing costs, $C_M$; and the reward obtained if the drug is a commercial success, $V_S$, or a commercial failure, $V_F$. For example, B may be of the form

$$B = V_S \cdot P_M + V_F(1 - P_M) - C_M$$

However, we may draw the decision tree for the research and development phase of the drug without regard to these details. It is given in Figure 30. By the technique of backward induction (see, e.g., Raiffa, 1968), we obtain the expected profit of this project:

$$BP_{S1}P_{S2}P_{S3} - C_1 - C_2P_{S1} - C_3P_{S1}P_{S2}$$

We have not yet discounted the value of future money. We may do so implicitly by requiring that B, $C_1$, $C_2$, and $C_3$ be valued in terms of present money.

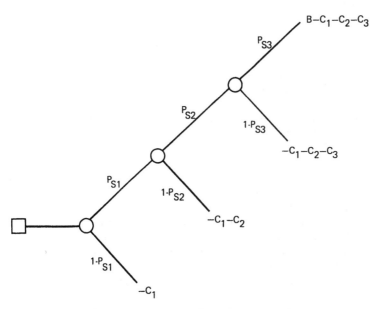

Figure 30 Decision tree for a three stage project.

The second important factor affecting profitability in a pharmaceutical research environment is the possibility of preemption by a competing firm. This may be taken into account by including the likelihood of preemption at each research phase. If $P_{Ei}$ denotes the conditional probability of preemption at time $t_i$, given that the project was not preempted at time $t_{i-1}$, we obtain

$$I' = BP_1P_2P_3 - C_1 - C_2P_1 - C_3P_1P_2$$

where $P_i = P_{Si}(1 - P_{Ei})$.

We may now obtain an index based on a benefit/cost ratio by dividing the expected net benefits by the expected research costs that will be charged against this and any future research and development budgets. This gives the project index

$$I_{Pearson} = \frac{BP_1P_2P_3 - C_1 - C_2P_1 - C_3P_1P_2}{C_1 + C_2P_1 + C_3P_1P_2}$$

It may be shown that selecting projects in descending order of this index will maximize the total expected net benefits (apart from possible end effects, which are discussed in Section 4.5) subject to a budgetary constraint on the total expect cost of the projects selected.

A different index may be based on Gittins's (1979) dynamic allocation index (DAI) for the optimal allocation of effort between a collection of alternative stochastic reward processes. These processes may in general be more complex than the research projects under present discussion. This index is sequential in the sense that it is recomputed at each review period in order to guide the reallocation of resources among projects as required by new information or changing circumstances. A theorem by Gittins shows that the DAI index is optimal for the following situation. Suppose that there are N projects, each divisible into stages, and that decisions as to which of the projects are to be worked on are taken as each stage is completed. If at each decision point one is permitted to choose just one project to continue for a further stage, the project with the highest DAI index should be chosen in order to maximize the total expected gain. Although this rather artificial restriction to working on just one project at a time is required for strict optimality, a policy that assigns priorities to projects according to their DAIs achieves results which are very close to optimality even when the restriction does not apply.

A theorem by Bergman (1982) shows that in a similar setting a modified form of the DAI is also optimal for choosing among alternative versions of a project. Roberts and Weitzman (1980) have also recently written on this topic. Less general results, aimed specifically at the research planning situation, are given by Glazebrook (1976,1978).

Two different but equivalent versions of the DAI for a reward process may be given. First, it may be characterized as the maximal rate of expected net (discounted) return per time unit that can be achieved by allowing the process to be continued up to an arbitrary stopping time which may depend on the current state of the process. For the previous example, with all three stages remaining to completion, the DAI is

$$I_{DAI} = \frac{BP_1P_2P_3 - C_1 - C_2P_1 - C_3P_1P_2}{1 - (1 - P_1)D_1 - P_1(1 - P_2)D_2 - P_1P_2D_3}$$

where $D_i$ is the present discounted value of one monetary unit at time $t_i$. As the discount rate approaches zero it may be shown that the DAI takes the form

$$I_{DAI} = \frac{BP_1P_2P_3 - C_1 - C_2P_1 - C_3P_1P_2}{t_1 + t_2P_1 + t_3P_1P_2}$$

An index that is equivalent to the DAI may be derived as follows. Construct the decision tree associated with the future of a (possibly ongoing) project and let $V(K)$ be its value if at any future time period the project may be retired (either through abandonment or completion) in favor of an alternative project with reward $K$ (suitably discounted). Let $K^*$ be the unique value satisfying the equation $K^* = V(K^*)$ [i.e., $K^*$ is the unique fixed point of $V(\cdot)$]. Then selecting the project with the largest fixed point $K^*$ is equivalent to selecting the project with the largest DAI. We may note that we have encountered a DAI of this form before, in Sections 2.6 and 3.2.

Of the profitability indices encountered in this section, one would anticipate the DAI index to be one of the most useful. Since it takes into account the sequential nature of the research process, it is clearly better than those discussed at the beginning of this section. One may also argue that in principle it will be better than the Pearson index, since the latter index does not, in the form given, take into account the value of freed resources when a project is abandoned.

As a final note on the topic of profitability indices, it should be observed that these indices do not take into consideration benefits of a nonfinancial nature. Hence it may often be worthwhile to complement such an index with a nonfinancial index. This may be done by splitting the checklist into two parts, one concerning the profitability of the project and the other detailing nonfinancial benefits. A scoring scheme may be adopted for the latter, while a profitability index is used for the former. The results for the various projects may be displayed as points on a $I_1 \times I_2$ graph, where $I_1$ is the horizontal nonfinancial scoring-index axis and $I_2$ is the vertical profitability-index axis. Unusual combinations of financial and nonfinancial benefits will then be highlighted.

## 4.5. INTEGER AND LINEAR PROGRAMMING

The techniques presented in the previous sections aim primarily at defining the relative merits of individual projects. That project selection based on rank procedures derived from these merit measures does not necessarily lead to optimal portfolios under resource constraints has been pointed out by several authors, including Bell (1969) and Gear (1974). Bell gives the following example. Assume that there are four projects under consideration, the benefits and costs of which are given in Table 12. The resulting benefit/cost ratios are then 1.25, 1.5, 2.5, and 1.75, respectively. Under the Olsen procedure, the projects are chosen in order of their benefit/cost rank until the wage and capital budgets are exhausted. If these budgets are 54 and 14, respectively, projects 3 and 4 are chosen and the total benefits for this portfolio sum to 58. An alternative portfolio could be obtained by selecting those that have the *worst* benefit/cost ratios. In this

Table 12  Project Details

|  | Project | | | |
|---|---|---|---|---|
|  | 1 | 2 | 3 | 4 |
| Wages cost | 24 | 30 | 10 | 4 |
| Capital cost | 4 | 10 | 2 | 12 |
| Benefit | 35 | 60 | 30 | 28 |

*Source*:  Courtesy of British Gas Council, OR Dept., London, England.

example we could choose projects 1 and 2, which together also satisfy the budget constraints. We then obtain a total benefit of 95, far exceeding the Olsen selection.

Although extreme, this example does demonstrate that determining the best portfolio under budget constraints is not equivalent to determining the relative merits of each project. With unyielding constraints and a fixed number of potential projects the portfolio selection problem can be seen to be similar to the problem of packing a suitcase in which we wish to maximize the value of the articles packed. Ranking procedures fail to achieve this maximum, for they fail to take into consideration the opportunity loss that occurs when available resources are not completely utilized.

Integer and linear programming formulations of the portfolio selection problem with given constraints have been extensively studied by Bell and Read (1970), and have been a principal feature of many project selection procedures (Freeman, 1960; Asher, 1962; Sobin, 1965; Nutt, 1965; Souder, 1967; Beged-Dov, 1965; Sengupta and Dean, 1960; Minkes and Samuels, 1966).

The simplest integer programming formulation of the portfolio problem has already been encountered in Section 4.4, where we discussed benefit/cost ratios. We assumed that each of n projects had associated with it a benefit and a cost and these were denoted by $B_i$ and $C_i$ for the ith project. In addition, we assumed that there was a total budget constraint of C. We may define the indicator variables $X_i$, i = 1,...,n, so that

$$X_i = \begin{cases} 0 & \text{if project i is not selected} \\ 1 & \text{if project i is selected} \end{cases}$$

The integer programming problem may then be defined as one of selecting $X_i = 0$ or 1 for i = 1,...,n so as to maximize the total benefits

$$\sum_{i=1}^{n} B_i X_i$$

subject to the budgetary constraint

$$\sum_{i=1}^{n} C_i X_i \leq C$$

This led to the benefit/cost ranking procedures when we did not concern ourselves with the possibility that the projects chosen did not exhaust all available resources. However, we now are concerned with these "end effects," for, as we have seen, they may completely alter the optimal solution. Hence we solve the problem above as given, utilizing any of several commercially available integer programming subroutines. The solution will specify which of the $X_i$ are 1 (select) and which are 0 (do not select), and the total value of the resulting portfolio.

The integer programming problem may be generalized in several different directions. First, we may identify several different types of resource constraint and incorporate these into the basic model. For example, there may be limitations on scientists, capital, and laboratory space. Denoting these limitations by S, C, and L, and the ith project's requirement for these resources by $S_i$, $C_i$, and $L_i$, we obtain the following problem:

$$\text{maximize} \sum_{i=1}^{n} B_i X_i$$

subject to the constraints

$$\sum_{i=1}^{n} S_i X_i \leqslant S$$

$$\sum_{i=1}^{n} C_i X_i \leqslant C$$

$$\sum_{i=1}^{n} L_i X_i \leqslant L$$

Second, project i may be available in any of $m_i$ different versions, differing, perhaps, in the amount of different resources to be assigned or the speed with which it is to be executed. To accommodate this generalization it is convenient to change our notation as follows. Let $X_{ij}$ take the value 1 or 0 according to whether the jth version of the ith project is selected or not. Correspondingly, let $B_{ij}$ denote the benefits resulting from selecting the jth version of the ith project. Analogous definitions hold for $S_{ij}$, $C_{ij}$, and $L_{ij}$. The required integer programming formulation of the problem is then

$$\text{maximize} \sum_{i=1}^{n} \sum_{j=1}^{m} B_{ij} X_{ij}$$

subject to the constraints

$$\sum_{i=1}^{n} \sum_{j=1}^{m_i} S_{ij} X_{ij} \leqslant S$$

$$\sum_{i=1}^{n} \sum_{j=1}^{m_i} C_{ij} X_{ij} \leqslant C$$

$$\sum_{i=1}^{n} \sum_{j=1}^{m_i} L_{ij} X_{ij} \leqslant L$$

$$\sum_{j=1}^{m_i} X_{ij} \leqslant 1, \quad i = 1, \ldots, n$$

The last set of constraints guarantee that at most one version of the ith project will be selected.

Third, compulsory or alternative compulsory projects may be incorporated into the program formulation. For example, if project 1 must be undertaken for reasons of prestige, this may be forced into the optimal solution by replacing the constraint

$$\sum_{j=1}^{m_1} X_{1j} \leqslant 1$$

by

$$\sum_{j=1}^{m_1} X_{1j} = 1$$

Similarly, if either project 1 and/or 2 must be included in any portfolio, this may be guaranteed by adding the constraint

$$\sum_{j=1}^{m_1} X_{1j} + \sum_{j=1}^{m_2} X_{2j} \geqslant 1$$

Fourth, different resources may be partially interchangeable. Biochemists may be able to do some of the work of biologists or chemists, and each of these may be able to perform some of the duties of a lab technician. To describe these possibilities, let

$s^1$, $s^2$, $s^3$, and T be resource constraints on biochemists, biologists, chemists, and lab technicians, respectively, and assume that we are able to transfer resources as shown in Figure 31. Let $s^{12}$ and $s^{13}$ denote the use of biochemists as biologists and chemists, respectively, and let $s^1$, $s^2$, and $s^3$ denote the use of biochemists, biologists, and chemists as technicians. We then obtain the program: Choose $X_{ij}$ and $s^1$, $s^2$, $s^3$, $s^{12}$, and $s^{13}$ to maximize

$$\sum_{i=1}^{n} \sum_{j=1}^{m} B_{ij} X_{ij}$$

subject to the constraints

$$\sum_{i=1}^{n} \sum_{j=1}^{m_i} S_{ij}^1 X_{ij} \leqslant s^1 - s^1 - s^{12} - s^{13}$$

$$\sum_{i=1}^{n} \sum_{j=1}^{m_i} S_{ij}^2 X_{ij} \leqslant s^2 - s^2 + s^{12}$$

$$\sum_{i=1}^{n} \sum_{j=1}^{m_i} S_{ij}^3 X_{ij} \leqslant s^3 - s^3 + s^{13}$$

$$\sum_{i=1}^{n} \sum_{j=1}^{m_i} T_{ij} X_{ij} \leqslant T + s^1 + s^2 + s^3$$

$$s^1 + s^{12} + s^{13} \leqslant s^1$$

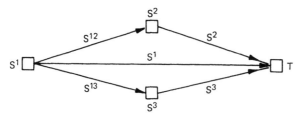

Figure 31 Constraints on use of personnel in non-specialist roles.

$$s^2 \leqslant S^2$$

$$s^3 \leqslant S^3$$

$$s^R \geqslant 0, \quad R = 1, 2, 3$$

$$\sum_{i=1}^{n} \sum_{j=1}^{m_i} C_{ij} X_{ij} \leqslant C$$

$$\sum_{i=1}^{n} \sum_{j=i}^{m_i} L_{ij} X_{ij} \leqslant L$$

We may note that this is no longer a zero-one integer programming problem, as $s^1$, $s^2$, $s^3$, $s^{12}$, and $s^{13}$ may take any integer value between 0 and the appropriate upper limit as specified above.

Fifth, resources may be earmarked for recruiting new scientific personnel. Similar to the technique used above, this is incorporated into the program by the introduction of dummy variables. In particular, let R denote a constraint on recruitment resources and let $r^k$ be recruitment of the kth type of scientific personnel. To obtain the corresponding integer programming problem, we substitute $S^k + r^k$ for $S^k$ in the first three constraints in the previous problem and add the constraints $\sum_{k=1}^{3} r^k \leqslant R$ and $r^k \geqslant 0$, $k = 1, 2, 3$.

Although seemingly complex, the formulations above are usually not adequate for any resource allocation problem. Implicitly, they assume that all resources necessary for a project are available at once, and hence do not take into consideration that a project usually lasts longer than one budgetary horizon, and faces a sequence of budgetary constraints over its life span. A naive approach to this problem would be to add an additional subscript t to each variable, where t defines the budget period, and to solve the corresponding integer programming.

The difficulty with this approach is that different projects or project versions have different lifetimes. As Bell and Reader point out, two projects having similar benefits, requiring the same total amount of each of the different type of resources but having different durations, may not be of equal merit. The shorter of the two will release resources sooner, and these can then be made available to other, perhaps presently unknown projects. On the other hand, it will also consume more resources initially. To get a handle on this new dimension, Bell introduces the concept of the opportunity value of freed resources. The technique is one of apportioning the estimated total benefits (per budgetary period) obtained from the research department among the resources available to it. Bell suggests that they be apportioned according to their marginal cost (i.e., all overheads are excluded). Assuming a certain consistency in the rate of effort per budgetary period, the total benefits per budgetary period can be extrapolated from past periods and assigned a value Q. Hence we obtain the following measure of opportunity value: Let $W(S)$ be the annual cost of a scientist, and similarly define $W(C)$ and $W(L)$ for capital and laboratory space, respectively. Then define the opportunity value of a scientist, $V(S)$, by the formula

$$V(S) = Q \frac{W(S)}{C \cdot W(C) + L \cdot W(L) + S \cdot W(S)}$$

To obtain a programming formulation we attach a subscript t to all variables, where $t = 1, \ldots, T$ and T is a planning horizon that is at least as long as the longest project under consideration. Now let $B_t(S)$ be the opportunity benefit of surplus scientists in budget period t. Similar definitions hold for $B_t(C)$ and $B_t(L)$. These may be computed as follows:

$$B_t(S) = V(S) \cdot s_t$$

$$B_t(C) = V(C) \cdot c_t$$

$$B_t(L) = V(L) \cdot l_t$$

where $s_t$, $c_t$, and $l_t$ are the surplus resource capacities in period t of scientists, capital, and laboratory space. The total value $B(S)$, $B(C)$, $B(L)$ of these surplus resources may now be obtained by discounting over individual budgetary periods $t = 1, \ldots, T$. For example, with an instantaneously compounded discount rate r, we obtain $B(S) = \sum_{t=1}^{T} B_t(S) e^{-rt}$. A first formulation of the problem then becomes

$$\text{maximize} \sum_{i=1}^{n} \sum_{j=1}^{m_i} B_{ij} X_{ij} + B(S) + B(C) + B(L)$$

[where $B(S)$, $B(C)$, and $B(L)$ are defined as indicated above] subject to the constraints

$$\sum_{i=1}^{n} \sum_{j=1}^{m_i} S_{ij} X_{ij} + s_t = S_t, \quad t = 1, \ldots, T$$

$$\sum_{i=1}^{n} \sum_{j=1}^{m_i} C_{ij} X_{ij} + c_t = C_t, \quad t = 1, \ldots, T$$

$$\sum_{i=1}^{n} \sum_{j=1}^{m_i} L_{ij} X_{ij} + l_t = L_t, \quad t = 1, \ldots, T$$

$$s_t \geq 0, \quad t = 1, \ldots, T$$

$$l_t \geq 0, \quad t = 1, \ldots, T$$

$$c_t \geq 0, \quad t = 1, \ldots, T$$

$$\sum_{j=1}^{m_i} X_{ij} \leq 1, \quad i = 1, \ldots, n$$

One weakness of this specific formulation is that it assigns a positive value to any resources not used in the first period. By definition such resources will not be used, and no benefits can be obtained from them. Bell and Reader suggest the following alteration. Define a budget period E at which we may expect surplus resources from the present projects to be profitably assigned to the new projects then available. Define for each resource, $V_t(\cdot)$, an opportunity value by linear interpolation: for example,

$$V_t(S) = \begin{cases} \dfrac{t}{E} V(S) & \text{for } 1 \leq t \leq E \\ V(S) & \text{for } t > E \end{cases}$$

The integer programming problem, written out completely for clarity, then becomes: Choose $X_{ij}$ equal to 0 or 1 *and* choose $s_t$, $c_t$, and $l_t$ for $t = 1, \ldots, T$ to maximize

$$\sum_{i=1}^{n} \sum_{j=1}^{m_i} B_{ij} X_{ij} + \sum_{t=1}^{T} V_t(S) s_t e^{-rt} + \sum_{t=1}^{T} V_t(C) c_t e^{-rt}$$

$$+ \sum_{t=1}^{T} V_t(L) l_t e^{-rt}$$

subject to the constraints

$$\sum_{i=1}^{n}\sum_{j=1}^{m_i} S_{ij}X_{ij} + s_t = S_t, \quad t = 1,\ldots,T$$

$$\sum_{i=1}^{n}\sum_{j=1}^{m_i} C_{ij}X_{ij} + c_t = C_t, \quad t = 1,\ldots,T$$

$$\sum_{i=1}^{n}\sum_{j=1}^{m_i} L_{ij}X_{ij} + l_t = L_t, \quad t = 1,\ldots,T$$

$$s_t \geqslant 0, \quad t = 1,\ldots,T$$

$$l_t \geqslant 0, \quad t = 1,\ldots,T$$

$$c_t \geqslant 0, \quad t = 1,\ldots,T$$

$$\sum_{j=1}^{m_i} X_{ij} \leqslant 1, \quad i = 1,\ldots,n$$

Additional features can be incorporated into the program in a rather straightforward manner. In general, we obtain either an integer programming problem or, as above, a mixed integer-continuous programming problem, where some of the variables (e.g., $c_t$ and $l_t$) are continuous variables. From a computational point of view, continuous variable programming problems are considerably easier to solve than integer variable problems. Several authors have suggested that all variables should, at least initially, be made continuous and the resulting problem be solved by a linear programming algorithm. There are several advantages to this approach. First, linear programming algorithms can efficiently handle large problems of several thousand variables and constraints. Second, the solution often contains many 0's and 1's for the $X_{ij}$'s. Third, when an $X_{ij}$ is neither 0 or 1, Gear and Lockett (1974) have found it to be the case that $\sum_{j=1}^{m_i} X_{ij} = 1$, thus suggesting that a new project version should be created. Fourth, linear programming algorithms yield additional useful information, such as the marginal increase that may be obtained in the objective function by a small increase in one of the resource constraints. This helps identify which resources could be most profitably increased or decreased.

## 4.6. STOCHASTIC PROGRAMMING

Although highly versatile, the integer and linear programming formulations outlined by Bell and Reader suffer one disadvantage that may make them inappropriate for exploratory research applications: They assume that once a project is selected it will be funded until completion. As pointed out in Section 4.4, this biases selection against projects where the initial stages of research are instrumental in determining, for a small expenditure of resources, the feasibility of successful completion.

Some methods have recently been proposed in the literature which recognize this important aspect of project selection. In this section we describe the stochastic programming formulation of Gear and Lockett (1973).

As in Section 4.4, we assume that it is possible to construct a decision tree for each project or project version. Figure 32 gives a simple example of a two-project selection problem requiring two types of resource inputs at each period. For each project a decision tree is drawn on a common time axis, with chance nodes indicated by circles and decision nodes by squares.

The technique of stochastic programming is one of expressing the multistage character of the sequential decision problem as an integer programming problem.

A sequence of increasingly more complex examples will reveal how this is done. Gear and Lockett may be consulted for additional details. Consider first a single resource problem in which there are only decision nodes (see Figure 33). Here there are two projects and three time periods. Project 1 may be carried out in one of two versions during time period 1, while project 2 does not split into two versions until time period 2. This is a totally deterministic problem and may be modeled as an integer programming problem in the following way:

$$\text{maximize } v_1 X_{11} + v_2 X_{12} + v_3 X_{21} + v_4 X_{22}$$

(where $X_{ij}$ is either 1 or 0 according as the project is selected or not selected for development) subject to the constraints

$$a_1 X_{11} + a_2 X_{12} + a_3 X_2 \leqslant A$$

$$b_1 X_{11} + b_2 X_{12} + b_3 X_{21} + b_4 X_{22} \leqslant B$$

$$c_1 X_{11} + c_2 X_{12} + c_3 X_{21} + c_4 X_{22} \leqslant C$$

$$X_{11} + X_{12} \leqslant 1$$

$$X_2 - X_{21} - X_{22} = 0$$

The quantities A, B, and C are resource constraints for periods 1, 2, and 3. The last constraint is a sequencing constraint; it ensures that if project 2 is selected in period 1, it is also continued in one of its versions at time period 2.

The next example (see Figure 34) introduces chance nodes, with these occurring at different time periods. By backward in-

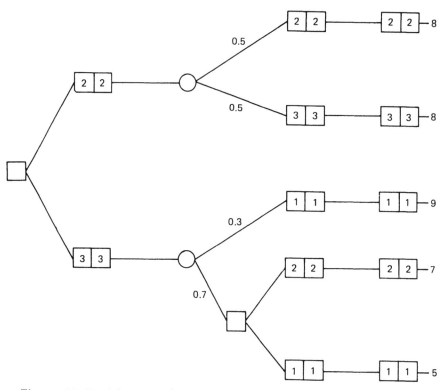

**Figure 32** Decision tree for two projects. For each stage of the project, a box denotes the amounts required of each of two types of resource. A square denotes a decision node and a circle a chance node. The probabilities associated with the chance node are written on the arms radiating out from the nodes. The final rewards upon completion are written in the last column.

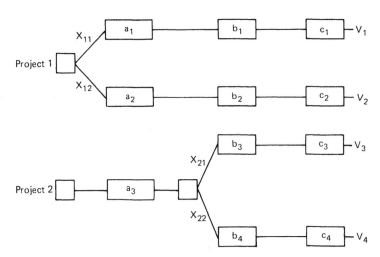

**Figure 33** Two projects with decision nodes.

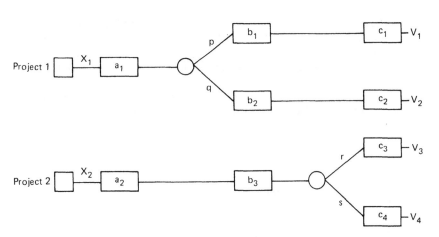

**Figure 34** Two projects with chance nodes.

Figures 32 through 36 on pages 210 through 214, © 1973 *IEEE Trans. Eng. Manage.*, Gear and Lockett (1973).

duction (see Raiffa, 1968, for details of this procedure) on each tree one obtains their respective values if chosen at time period 1. Hence the problem is one of maximizing

$$(pv_1 + qv_2)X_1 + (rv_3 + sv_4)X_2$$

Our constraints will depend on what happens at the chance nodes. Gear and Lockett require that whatever happens, the demand for resources not exceed the supply. There are two possible outcomes for time period 2 and four for time period 3. Hence we obtain the constraints

$$a_1X_1 + a_2X_2 \leqslant A \quad \text{(period 1)}$$

$$\left.\begin{array}{l} b_1X_1 + b_3X_2 \leqslant B \\[2mm] b_2X_1 + b_3X_2 \leqslant B \end{array}\right\} \text{(period 2)}$$

$$\left.\begin{array}{l} c_1X_1 + c_3X_2 \leqslant C \\[2mm] c_1X_1 + c_4X_2 \leqslant C \\[2mm] c_2X_1 + c_3X_2 \leqslant C \\[2mm] c_2X_1 + c_4X_2 \leqslant C \end{array}\right\} \text{(period 3)}$$

Decision nodes and chance nodes may be combined into a single problem as is done in Figure 35. Here the decision of which version of project 1 to choose at time period 2 is conditional on the outcome of the chance node for project 2. Thus $X_{11}$ will be written as either $X_{11}^p$ or $X_{11}^q$, depending on whether the upper branch (marked p) or the lower branch occurred for project 2. By backward induction we obtain the objective function

$$(pv_3 + qv_4)X_2 + p(v_1X_{11}^p + v_2X_{12}^p) + q(v_1X_{11}^q + v_2X_{12}^q)$$

subject to the constraints

$$a_1X_1 + a_2X_2 \leqslant A \quad \text{(period 1)}$$

$$\left.\begin{array}{l} b_3X_2 + b_1X_{11}^p + b_2X_{12}^p \leqslant B \\[2mm] b_4X_2 + b_1X_{11}^q + b_2X_{12}^q \leqslant B \end{array}\right\} \text{(period 2)}$$

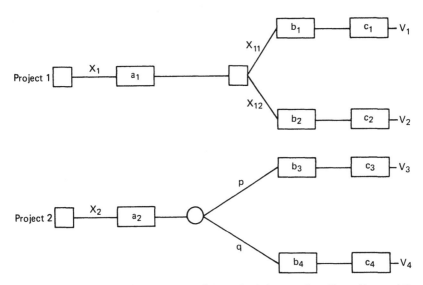

**Figure 35** Two projects, one with a decision node, the other with a chance node.

$$\left.\begin{array}{l} c_3 X_2 + c_1 X^p_{11} + c_2 X^p_{12} \leqslant C \\[2em] c_4 X_2 + c_1 X^q_{11} + c_2 X^q_{12} \leqslant C \end{array}\right\} \text{(period 3)}$$

$$X_1 - X^p_{11} - X^p_{12} = 0$$

$$X_1 - X^q_{11} - X^q_{12} = 0$$

As in problem 1, the last two constraints specify that project 1 will be continued in some version in time period 2 if it is selected initially.

As a final example, consider Figure 36. This problem involves chance followed by decision branching on project 1, and the reverse on project 2. The integer programming formulation is to maximize

$$pv_1 X_{11} + pv_2 X_{12} + qv_3 X_1 + v_4 X_{21} + (rv_5 + sv_6) X_{22}$$

subject to the constraints

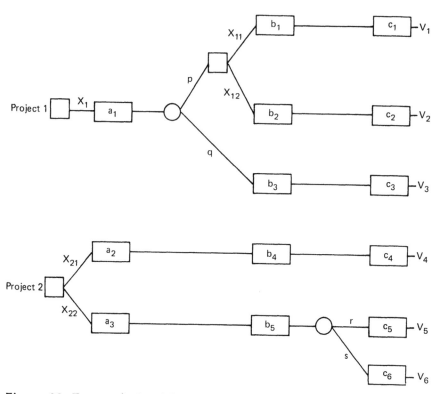

**Figure 36** Two projects with both decision and chance nodes.

$$a_1 X_1 + a_2 X_{21} + a_3 X_{22} \leqslant A \quad \text{(period 1)}$$

$$\left.\begin{array}{l} b_4 X_{21} + b_5 X_{22} + b_1 X_{11} + b_2 X_{12} \leqslant B \\[2mm] b_3 X_1 + b_4 X_{21} + b_5 X_{22} \leqslant B \end{array}\right\} \text{(period 2)}$$

$$\left.\begin{array}{l} c_4 X_{21} + c_5 X_{22} + c_1 X_{11} + c_2 X_{12} \leqslant C \\[2mm] c_4 X_{21} + c_6 X_{22} + c_1 X_{11} + c_2 X_{12} \leqslant C \\[2mm] c_3 X_1 + c_4 X_{21} + c_5 X_{22} \leqslant C \\[2mm] c_3 X_1 + c_4 X_{21} + c_6 X_{22} \leqslant C \end{array}\right\} \text{(period 3)}$$

$$X_{21} + X_{22} \leqslant 1$$

$$X_1 - X_{11} - X_{12} = 0$$

In general we may note that for each time period there are as many constraints as there are possible outcomes realizable in that time period. This may lead to very large integer programming problems. Gear and Lockett computed the optimal plan for a six-project selection problem with 30 variables and 29 constraint rows. This took 12 seconds of computer time using a modified Balas Algorithm on an ICL 1906A computer. Unfortunately, they also found that the continuous linear programming formulation of this problem, which as usual took much less computer time, did not yield comparable results. This implies that there is little hope at the present that large problems can be solved inexpensively.

One method of reducing the complexity of the stochastic programming formulation of the selection problem is to eliminate decision nodes for any projects that depend on outcomes relating to another project. Then the decision tree for each project may be reduced by backward induction to a tree containing only chance nodes, and there is just one indicator variable $X_i$ for each project i.

Another serious drawback of stochastic programming that needs to be overcome is its requirement that *all* outcomes of a portfolio be forced to lie within the budget constraint. This is requiring too much. In the first place, this would be guarding against some possibly very unlikely events. In the second, if such an unlikely event were the appearance of several very promising compounds on the horizon at once, one would hope that additional funding could be obtained to explore them simultaneously. One method of handling this second type of eventuality would be to adjust the corresponding constraint in an explicit fashion, [e.g., in (*) on page 212 replace $b_2X_1 + b_3X_2 \leqslant B$ with $b_2X_1 + b_3X \leqslant 2B$] The former could be catered for by replacing a constraint on all outcomes by a constraint on the average outcome. For example, if the benefit $V_i$ and cost $C_i$ of project i are random, then one could pose the integer programming problem

$$\text{maximize} \sum_{i=1}^{n} X_i E(V_i)$$

subject to the constraints

$$\sum_{i=1}^{n} X_i E(C_i) \leqslant C$$

where $E(\cdot)$ denotes the expectation of the random variable. This formulation of the problem yields Pearson's index when end effects are not taken into account. If one wanted to guard against more than just the average future exceeding the budget, one could replace the constraint above by, say,

$$\sum_{i=1}^{n} X_i [E(C_i) + \delta \text{ Var}(C_i)] \leqslant C$$

where $\text{Var}(C_i)$ is the variance of $C_i$, and $\delta$ is some appropriately chosen nonnegative constant. Following the techniques of the preceding section, this basic model could be elaborated in several directions. Furthermore, it may well be that the linear programming version of this problem would give results close to that of the original integer programming formulation. A rather similar approach is discussed in the next section.

In summary, for small problems the detailed stochastic programming formulation given by Gear and Lockett could be useful. However, for a pharmaceutical research environment a model of greater flexibility for handling uncertainty would be desirable.

## 4.7. OTHER INTEGER AND LINEAR PROGRAMMING TECHNIQUES FOR HANDLING UNCERTAINTY

Lockett and Gear (1972) have suggested that one method of overcoming the dimensionality difficulties encountered in stochastic programming is the use of simulation. The technique may be illustrated by reviewing portions of a model proposed by Allen and Johnson (1970).*

Like many other authors, they point out that the benefits $B_{ij}$ to be received from completing version j of project i usually can-

---

*The notion of credibility used by Allen and Johnson is not equivalent to a subjective probability distribution over the space of possible outcomes. Our interest here is not, however, in the usefulness of this concept, only in their use of simulation. It may be assumed that we are dealing with probability distributions.

not be known with certainty. However, one may identify a credible range of values that $B_{ij}$ may take. Rather than replacing this credible range with a single-point estimator representing either the most probable or the expected value of $B_{ij}$, they suggest that a sequence of linear programming problems be generated from the set of possible $B_{ij}$ values. This may be done by randomly sampling the distribution of $B_{ij}$ for each i and j, where each project i is taken to have an outcome independent of any other distinct project k, but where similar versions of a project are assumed to experience similar outcomes. For each set of outcomes $B_{ij}$, $i = 1, \ldots, n$; $j = 1, \ldots, m_i$, a linear program is run and the best portfolio is identified. Proceeding in this way, we can identify a collection of portfolios, many of which will be optimal for several different sets of outcomes. Each portfolio is then evaluated for each of these sets of outcomes, and a histogram of benefits is obtained. Allen and Johnson ran 39 sets of possible outcomes and found one portfolio to be optimal in seven cases. This portfolio, as well as those others which were optimal in more than one case, also performed well for most of the 39 sets of outcomes. Any of these portfolios could reasonably have been chosen.

There are several advantages of the Allen and Johnson method. First, it automatically yields a sensitivity analysis with respect to the data. Second, it identifies several nearly optimal portfolios, which will permit the user to select one on the basis of criteria not modeled in the program. Third, each integer programming problem may be approximated by a linear programming formulation that will accommodate large problems. Fourth, it permits other uncertainties, such as uncertainties in future budget constraints, to be modeled as well.

A different approach to uncertainty is chance-constrained programming. This technique was pioneered by A. Charnes and W. W. Cooper (1963) and has been applied to some project selection problems by Watters.

The principle here is one of maximizing the expected reward among portfolios, subject to probabilistic constraints on resource usage. Gear et al. (1971) give the following description of Watters' dissertation. Assume that there are n independent projects and T time periods. For each project i and time period t identify a cost $\tilde{C}_{it}$, with expectation $C_{it}$ and variance $\sigma^2_{C_{it}}$, and a reward $\tilde{r}_{it}$, with expectation $r_{it}$ and variance $\sigma^2_{r_{it}}$. Thus $\tilde{C}_{it}$ and $\tilde{r}_{it}$ are random variables, and for any portfolio $(X_1, X_2, \ldots, X_n)$ the cost and reward at time period t are the random variables

$$\tilde{C}_t(X_1,\ldots,X_n) = \sum_{i=1}^{n} \tilde{C}_{it} X_i \quad \text{and} \quad \tilde{r}_t(X_1,\ldots,X_n) = \sum_{i=1}^{n} \tilde{r}_{it} X_i$$

The chance-constrained property of the formulation specifies that the cost at time t should be no greater than some prespecified amount $B_t$ with probability $\alpha_t$, that is,

$$\text{Prob}(\tilde{C}_t(X_1,\ldots,X_n) > B_t) \leqslant \alpha_t$$

We may note that the expectation and variance of $\tilde{C}_t(X_1,\ldots,X_n)$ are

$$E(\tilde{C}_t) = \sum_{i=1}^{n} C_{it} X_i \quad \text{and} \quad \sigma^2(\tilde{C}_t) = \sum_{i=1}^{n} \sigma^2_{C_{it}} X_i$$

Hence, if we assume that $\tilde{C}_{it}$ is normally distributed, we obtain

$$\alpha_t \geqslant \text{Prob}(\tilde{C}_t > B_t) = \text{Prob}\left( \frac{\tilde{C}_t - E(\tilde{C}_t)}{\sigma(\tilde{C}_t)} > \frac{B_t - E(\tilde{C}_t)}{\sigma(\tilde{C}_t)} \right)$$

or

$$1 - \alpha_t \leqslant \Phi\left( \frac{B_t - E(\tilde{C}_t)}{\sigma(\tilde{C}_t)} \right)$$

where $\Phi(\cdot)$ is the standard normal distribution function. Using tables for this function, we may identify the quantity $Z_{\alpha_t}$ satisfying $1 - \alpha_t = \Phi(Z_{\alpha_t})$. Thus our chance constrained inequality is essentially

$$\frac{B_t - E(\tilde{C}_t)}{\sigma(\tilde{C}_t)} \geqslant Z_{\alpha_t}$$

This is equivalent to a quadratic constraint containing terms of the form $C_{it} C_{jt} X_i X_j$. Although not mentioned by Gear et al., observe that by setting $X_{ij} = X_i \cdot X_j$, $C_{ijt} = C_{it} \cdot C_{jt}$, and introducing constraints $2X_{ij} - X_i - X_j = 0$, this reduces to a zero-one integer programming problem with $(n^2/2) + n$ variables.

Rather than specifying the objective function to be the expected reward $\Sigma_{i=1}^{n} (\Sigma_{t=1}^{T} r_{it}) X_i$, Watters suggests that a risk-avoidance utility-like measure be used instead. In particular, he proposes that the function to be maximized should be

$$\sum_{i=1}^{n} (\mu_i - K\sigma_i^2)X_i$$

where $\mu_i$ and $\sigma_i^2$ are the mean and variance of the net return of preoject i. That is,

$$\mu_i = \sum_{t=1}^{T} (r_{it} - C_{it})$$

$$\sigma_i^2 = \sum_{t=1}^{T} (\sigma_{r_{it}}^2 + \sigma_{C_{it}}^2)$$

Watters treats the constant K as an unknown, preferring to generate the complete set of "optimal" solutions for all positive values of K. It is suggested that the decision maker can then direct his or her attention to the relative merits of the portfolios in this greatly reduced set.

The direct applicability of chance-constrained programming to pharmaceutical research is doubtful. However, there are several interesting and potentially useful ideas contained in this formulation. In particular, we may note that whereas stochastic programming requires that *all* outcomes of a portfolio be forced to lie within the budget constraints, chance-constrained programming requires only that most of them do.

Another interesting aspect of the Watters formulation is his specification of a risk-avoidance utility-like function. On the other hand, an unusual feature of the specific formulation he uses is that projects with both long upper distribution tails and short lower tails are penalized. (Of course, this possibility does not fall within his assumption of normality, but such an assumption may not hold in general.) Alternatively, one may introduce risk avoidance as a constraint of the form, say,

$$P\left(\sum_{i=1}^{n} r_{it}X_i \leqslant R_t\right) \leqslant B_t$$

which guarantees that for any portfolio selected, the probability of a return less than $R_t$ in period t is less than $B_t$. Simulation may be used to see how the optimal portfolio varies with $B_t$.

The primary weakness of chance-constrained programming is that it assumes independence between time periods. This is not really appropriate. A large value for $C_{it}$ in one time period implies that something has happened which may affect future costs.

For example, a cost of zero in term period $C_{it}$ may correspond to the possibility that the project has been abandoned. Hence all future costs should also be zero.

## 4.8. DYNAMIC PROGRAMMING AND NONLINEAR MODELS

An alternative to stochastic programming is dynamic programming, a powerful analytic approach that easily accommodates sequential decision problems. Dynamic programming is based on Bellman's principle of optimality (1957), which states: "An optimal policy has the property that whatever the initial state and decision are, the remaining decisions must constitute an optimal policy with regard to the state resulting from the first decision." This principle may be formalized as follows. Let $X_t$ be the state at time t, let $d_t$ be the state at time t + 1 which results from taking decision $d_t$ at time t, and let $f_t(X_t)$ be the value of the optimal policy from time period t onwards, given that we are presently at state $X_t$. Then, in a deterministic setting, the optimality principle is equivalent to the recursive relationship

$$f_{t-1}(X_{t-1}) = \max_{d_{t-1}} f_t(X_t(X_{t-1}, d_{t-1}))$$

In particular, if we knew the function $f_t$, we would know which decision $d_{t-1}$ to choose at time t − 1, and hence be able to calculate the function $f_{t-1}$. This leads to a process of iteration backward in time when the time horizon is finite. At time T, the final time period, we will know what the optimal decision is since we need only optimize with respect to the immediate time period before us. Having established $f_T$ we compute $f_{T-1}$, and then $f_{T-2}$, and recursively backward up to $f_1$. Simultaneously, we derive the optimal decision function $d_T^*$, $d_{T-1}^*$, ..., $d_1^*$, which defines the optimal policy.

One advantage of dynamic programming over stochastic programming is that the functional form of f may be completely general and not just linear. Two well-known models using dynamic programming for project selection are those by Hess (1962) and Rosen and Souder (1965). These will be discussed at the beginning of this section. An alternative nondynamic programming formulation of a model similar to theirs will be discussed at the end.

In the Hess model it is assumed that research and development of a selected project is continued until one of the following

two milestones is reached: The project achieves "technical success," or a predetermined time horizon is reached without success, at which time the project is terminated as unsuccessful.
The benefits of a successful project are defined by the values $G_t$,
$t = 1, \ldots, T$, where $G_t$ is defined to be the total expected discounted gross profit, given that the project is successful in the
t'th period and then commercialized. Generally, $G_t$ will be a decreasing function of t, reflecting the value of obtaining success
earlier rather than later. Hess recognizes that the probability of
success in any time period t is a function of the amount of resources devoted to the project. Hence the probability of success
in time period t is defined to be a function $P_t(x_t, y_t)$ of the
amount of resources $x_t$ assigned to the project in time period t
and the amount of resources $y_t$ previously assigned to the project
before time period t. Using the optimality principle, the following
recursive relationship is derived for the monetary value of a
project:

$$f_t(y_t) = \max_{x_t \geqslant 0} \{G_t P_t(x_t, y_t) - x_t + \rho[1 - P_t(x_t, y_t)]f_{t+1}(y_{t+1})\}$$

where $f_t(y_t)$ is the expected value of a project at time t, given
that $y_t$ resources have already been assigned and $\rho$ is the discount factor. The backward induction may be initiated at time T
by the relationship

$$f_T(y_T) = \max_{x_T \geqslant 0} \{G_T P_T(x_T, y_T) - x_T\}$$

If the present time is t = 1, the present value of the project is
$f_1(0)$.

Computing $f_t(y_t)$ for any given t and $y_t$ may be difficult.
Hess suggests two forms which are analytically tractable. In the
first he suggests

$$P_t(x_t, y_t) = 1 - \exp(-C_t x_t)$$

where $C_t$ is chosen to be the reciprocal of the investment necessary to achieve 0.63 probability of technical success. We note
that in this form the probability of success in time period t is independent of past effort $y_t$. Hence lack of success in previous
periods implies that one is "starting from scratch" in the present
period. Hess claims that this model may be applicable to drug
screening, and in some instances this may be true. The recursion formula becomes

$$f_t = \max_{x_t \geqslant 0} \{G_t[1 - \exp(-C_t x_t)] - x_t + \rho \exp(-C_t x_t)f_{t+1}\}$$

and by induction one obtains

$$x_t^0 = \begin{cases} \dfrac{1}{C_t} \ln[C_t(G_t - \rho f_{t+1})] & \text{if } C_t(G_t - \rho f_{t+1}) > 1 \\[2mm] 0 & \text{otherwise} \end{cases}$$

$$f_t = G_t - \frac{1}{C_t} - x_t^0 = G_t - \frac{1}{C_t}\{1 + \ln[C_t(G_t - \rho f_t]\} \quad \text{if } x_t^0 \neq 0$$

where $x_t^0$ is the $x_t$ that gives the maximum in the recursion relationship. By backward induction the values of $x_t^0$ and $f_t$ may be
obtained for any $C_1,\ldots,C_T$ and $G_1,\ldots,G_T$.

In the second model, Hess permits current success to depend
on past effort. He assumes that the probability of success in the
t'th period is

$$P_t(x_t,y_t) = 1 - \exp(-Cy_t x_t)$$

where C is a constant and $y_t$ is a weighted sum of past research
efforts. In particular, he sets

$$y_t = \sum_{s=1}^{t-1} w_s x_s$$

where the weights $w_s$ are chosen so that effort at time s is more
valuable for current success than effort at time $s - 1$ (i.e.,
$w_s > w_{s-1}$). A special form of these weights is obtained by
letting $w_s = w^{t-s}$ for $y_t$. It is then readily shown that

$$y_t = w(y_{t-1} + x_{t-1})$$

Furthermore, if $G_t = G$, for all t, we obtain the recursion

$$f_T(y_T) = \max_{x_T \geqslant 0} \{G[1 - \exp(-Cy_T x_T)] - x_T\}$$

$$f_t(y_t) = \max_{x_t \geqslant 0} \{G[1 - \exp(-Cy_t x_t)] - x_t$$

$$+ \exp(-Cy_t x_t)\rho f_{t+1}(w(y_t + x_t))\} \qquad t < T$$

By solving this equation by backward induction we may obtain the optimal allocation $x_1^0, \ldots, x_T^0$ for the project.

We should carefully note what has been solved thus far. The problem that Hess has solved above is how resources should be allocated to a *single* project over time *without* budgetary constraints. He is identifying the most profitable *version* of a project. In contrast to the linear programming formulations of the same problem, it is noteworthy that increasing the amount of resources assigned to a project does not necessarily increase its profitability.

Hess now goes further and considers the allocation of resources among projects given a budgetary constraint in the first period. By means of a Taylor series expansion, he obtains the following interesting policy for his first model: Work on those projects that would have the highest probabilities of success under conditions of optimal budgeting and no aggregate budgetary constraint. The cutoff point and the exact amount assigned to each selected project are more complicated, and the exact formulas may be found in Hess (1962).

A weakness of the Hess result is that a budgetary constraint is given only for the first period; for all periods thereafter unlimited funds are assumed. We may note, however, that when there is no budgetary constraint, then for his first model the maximum expenditure for a project is given in the first time period. This suggests that the inclusion of additional budgetary constraints for other time periods equal in magnitude to the first period constraint may not greatly alter the optimal allocation.

Rosen and Souder offer a modification of the Hess model. The basic recursion relationship for a single project is simplified to

$$f_t(y_t) = \max_{x_t} \{GP(x_t) - x + \rho[1 - P(x_t)]f_{t+1}(x_t + y_t)\}$$

where $P(x)$ does not depend on y, and where G and P are independent of the time period t. Additional constraints are, however, imposed on $x_t$. Specifically, if $S_{max}$ and $S_{min}$ are the maximum and minimum total amounts that management believes can be spent on a project during its life, and $X_{max}$ is the maximum that can be spent in any one time period, we have the restrictions

$$\min\{X_{max}, S_{max} - y_t\} \geq x_t \geq \max\{\min(X_{max}, S_{min} - y_t), 0\}$$

on $x_t$ for any time period t. These restrictions do not make it any more difficult to compute the expected value function $f_t(y_t)$

numerically, and permit management to have greater control over the type of policies they are willing to consider.

Like Hess, Rosen and Souder permit G and P to differ for different projects, and notationally they are distinguished from each other by adding a superscript for the project number. An overall budgetary constraint is introduced for the first period of the form

$$\sum_{i=1}^{N} x_i \leq B$$

where N is the number of projects under consideration. Unlike Hess, however, the portfolio problem is now solved by applying dynamic programming techniques across the first period value functions $f_1^i(0)$. Specifically, set

$$H^1(b) = \max_x f_1^1(x), \quad 0 \leq b \leq B$$

where the maximum is taken over the set

$$\min\{S_{max}^1, X_{max}^1, b\} \geq x \geq \min\{S_{min}^1, X_{max}^1\}$$

and then recursively define

$$H^i(b) = \max_x \{f_1^i(x) + H^{i-1}(b - x)\}$$

where the maximum is now taken over

$$\min\{S_{max}^1, X_{max}^1, b\} \geq x \geq \min\{S_{min}^1, X_{max}^1\}$$

Setting b = B for the argument in $H^N(\cdot)$, one obtains the value of the optimal policy, and by backward induction one may also obtain the optimal allocation between projects. The best unrestrained allocations in future time periods for the selected projects may then be determined from the function $f_t^i$, t = 2,...,T, depending on the initial first period allocation.

Rosen and Souder do not consider any specific form for the probability functions $P^i(x)$, where i ranges over the projects. Instead, $P^i(x)$ is assumed known at the points $\{x_j, j = 1,2,...,n\}$, and intermediate values are interpolated. Some typical curves for $P^i(x)$ are given in Figure 37. This technique gives the Rosen-Souder model considerable flexibility. It may, of course, be applied to other models as well.

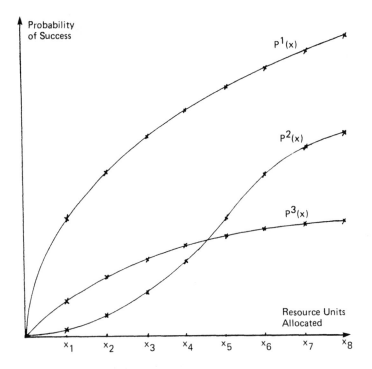

**Figure 37** Typical relationship between success probabilities and effort. (©1965 *IEEE Trans. Eng. Manage.*, Rosen and Souder (1965).

A weakness of both the Hess and Rosen-Souder models is that they do not permit more than one type of budgetary constraint, and the one constraint they do permit is limited to just one time period. A very interesting paper by Atkinson and Bobis (1969) overcomes part of this difficulty. In addition, several other features of the selection problems are highlighted. To achieve their results, dynamic programming is not used; rather, they show how a constrained nonlinear programming problem may be partly transformed into an unconstrained problem. This permits standard hill-climbing techniques to be used for discovering the optimal allocation.

Like Hess, and Rosen and Souder, Atkinson and Bobis emphasize the problem of achieving an optimal allocation of resources within a single project with respect to a sequence of budgets encompassing its life, as well as that of achieving an optimal allocation of resources between various competing projects. They see the former problem as one of balancing two conflicting costs. First, delaying the completion of a project will severely influence its market value when it is completed. Second, rushing a project will necessarily lead to inefficient use of research resources. We shall consider each in turn.

The volume of sales in year j if the project is completed in year 1 is assumed to grow according to the form

$$\frac{B}{1 + \exp(\gamma - \rho j)} = g(j)$$

where $B$, $\gamma$, and $\rho$ are constants chosen to reflect the marketing prospects of the project. This logistic curve increases slowly initially, and then more rapidly, before reaching a plateau level. Realistically, sales should also decrease after a peak period. This feature is incorporated by simply setting $g(j) = 0$ for all sufficiently large j (e.g., for $j \geqslant 11$).

A delay of several years may be required to prepare fully the product that results from the project for the market. Denote this by d years. Atkinson and Bobis suggest the following form for the sales volume in year j of a product whose research is completed in year i:

$$S_{ij} = k^n g(j - n)$$

where $n = i + d - 1$ and $k$ is an appropriately chosen constant. Thus the sales volume decreases exponentially with time.

The sales are conditional on completion of the research project in year i, followed by successful commercialization. The probabilities of these two respective events are denoted by $c_i$ and $P$, and the probability of their joint occurrence is assumed to be $p_i = c_i P$. Hence, in their model the probability of commercialization is independent of when the project is completed. Furthermore, project completion is equivalent to successful completion of the research phase of the project. A restriction of five years is placed on the permissible time to completion. Noncompletion within that time period implies that the project is abandoned.

The probability of completion, $c_i$, in year i is, of course, partly a function of the amount of resources assigned to the project prior to and including period i. It is also a function of how effectively these resources are distributed over these time periods.

Atkinson and Bobis model these two aspects in the following way. It is assumed that there exists a sequence $y_i$, $i = 1, \ldots, 5$, of maximal efficient allocations over the life of the project, such that if $x_i$ is the true allocation in period i, the effective allocation is

$$z_i = \begin{cases} x_i & \text{if } x_i < y_i \\ y_i \left(\dfrac{x_i}{y_i}\right)^{\varepsilon_i} & \text{if } x_i > y_i \end{cases}$$

The exponent $\varepsilon_i$ is chosen to be less than 1, so that allocations above $y_i$ exhibit decreasing marginal returns. The sequence $y_i$, $i = 1, \ldots, 5$, necessarily depends on allocations made in the preceding years, since a large increase in $x_i$ in one year implies that effort has to be spent in the next training new staff. To capture this nuance, suppose that $y_i^0$ is the original estimate of the efficient expenditure in year i. Then set

$$y_i = y_i^0 \left[ 1 + \frac{1}{2}\left(\frac{z_{i-1}}{y_{i-1}^0} - 1\right)\right]$$

The $y_i$ values now increase or decrease depending on the expenditure in the preceding year.

The probability of completion in year i is now also defined in terms of the effective allocation $z_i$. Specifically, let $w_i = \Sigma_{j=1}^{i} z_j$ and then the cumulative distribution function for completion by time i for total effort $w_i$ is set equal to

$$C_i = C(w_i) = \frac{1}{1 + \exp(\alpha - \beta w_i \gamma)}$$

where $\alpha$, $\beta$, and $\gamma$ are appropriately chosen constants. Thus

$$c_i = C_i - C_{i-1}$$

The cumulative probability of successful commercialization by year i is therefore,

$$P_i = C_i P$$

The return function used by Atkinson and Bobis to express the net benefit of a project is similar to those used by other researchers. It consists of two parts. First, the net revenue accruing to the company from sales in year j as a result of a research success in year i is

$$G_{ij} = S_{ij}[Q_j(1 - H) - M_j]\left(\frac{1}{1 + D}\right)^{j-1}$$

where $Q_j$ is the selling price in year j, H an overhead cost factor, $M_j$ the manufacturing cost, and D a discount constant. This expression is essentially the standard accounting formula for net revenue: revenue = sales volume × unit profit per sale × discount factor. Summing over the life of the product gives the payoff from a success in year i:

$$G_i = \sum_{j=i+1}^{J} G_{ij}$$

The expected payoff is therefore

$$G = \sum_{i=1}^{5} p_i G_i$$

Second, the expected cost of research R is the sum of the expected research expenditures $R_i$ in each of the years $i = 1, \ldots, 5$, suitably discounted:

$$R = \sum_{i=1}^{5} R_i \left(\frac{1}{1+D}\right)^{i-1}$$

Each $R_i$ is obtained by multiplying the allocated ith-year budget $x_i$ by the probability $(1 - C_{i-1})$ that the project is not completed before the beginning of the ith year.

Combining the two factors above, one obtains the expected net benefit

$$V = G - R$$

While the above defines the reward and its dependence on the budget pattern, the optimal allocation remains to be found. If there are L projects, a typical budget pattern may be represented by a L × 5 matrix X, where the (l,m)th element is the amount of money to be spent on the lth project in the mth year if it has not been completed by that time. The overall budget constraint on resources will be satisfied by some of these budget patterns. If the set of budgets that satisfy it is denoted by T, the optimal allocation has the value

$$\sup_{x \in T} \sum_{i=1}^{L} V_l$$

where $V_l$ is the value of the lth project under budget X. Atkinson and Bobis show how this problem may be restructured into a nonconstrained optimization problem and solved by an iterative routine.

## 4.9. RESPRO: AN INTERACTIVE MODEL

Almost all the programming models we have discussed so far seek
to find one optimal allocation among the set of all allocations. A
major deficiency in this approach, one that may partly explain
management's reported lack of interest in the presently available
project selection methodologies, is that they do not permit a research manager to participate in the selection process. Any
model, even one specifically designed for a particular environment, cannot incorporate all the elements that will be important
in any one specific allocation problem. Hence, almost by definition, a good project allocation methodology is one that permits the
research manager to control the optimization procedure in such a
way that these unique, unprogrammable elements may be brought
into consideration as the need arises. Such a process is interactive and the need for interactive procedures is becoming increasingly recognized.

Gittins and Roberts (1981) have devised an interactive computer-based model suitable for pharmaceutical selection and allocation problems. So far, one company has purchased a license to
use it. For any given allocation plan the program will determine
a collection of parameters that will describe the financial and
performance implications of the plan. Included among these financial indicators are (1) the expected total present value of cash
flows resulting from the selected projects, excluding research
costs, and the contribution that each project makes to this total;
(2) the expected total discounted effort that is required by these
projects, together with the contribution of each project to this
total; and (3) the expected present value generated per unit of
discounted effort allocated to existing projects. For each project
and each budgetary planning period, the following performance
characteristics are given: (1) the probability of completion by
that period, (2) the probability of preemption by that period,
and (3) the probability that work is still in progress at the end
of that period. These indicators may be used to determine the
risk associated with the allocation plan and the probable availability of resources for other projects at each time period in the
future.

Other financial parameters indicate in a systematic way how
the initial allocation plan may be improved. Most important among
these is the marginal profitability index. This quantity, whose
value varies according to both the project and period under discussion, is the rate at which the expected value of the research
program increases as small increases are made to the expected

discounted effort allocated to the project at the beginning of the period concerned. This index helps identify those projects whose allocations could profitably be increased, and those that should be cut back or perhaps eliminated altogether. A similar index calculates the overall expected marginal present value per unit increase in discounted effort for the research program as a whole. Here small increases in expected discounted effort are distributed over all projects and planning periods in amounts that are proportional to the prevailing distribution of effort among projects and periods. This quantity may be compared with the marginal cost per unit increase in the total discounted effort allocated to the research program as an indication of whether the program should perhaps be expanded or contracted. A final parameter indicates the average discount rate.

The Gittins-Roberts model may be used further to compare different allocations in a systematic way. This permits the research manager to answer "what if" questions relating to changes in the input parameters, and to compare allocations that fulfil certain unmodeled considerations with those that do not.

While drawing the research manager into active participation in finding a good, nearly optimal allocation, the Gittins-Roberts approach to project slection also permits realistic and significantly complex relationships to be modeled. This will now become evident as the assumptions and details of the program are described.

First, we shall model the financial benefits of a single project. Like Atkinson and Bobis, Gittins and Roberts assume that the value of a project is a function of its time of completion. Hence, let $V(t)$ denote the present value of the cash flows generated by the project if it is successfully completed at time $t$. The probability of success by time $t$ is, of course, a function of the amount of effort by time $t$. Denote, therefore, the rate of effort at time $t$ by $u(t)$. The integral $U(t) = \int_0^t u(s)\, ds$ is then the total effort expended by time $t$. Again, like Atkinson and Bobis, Gittins and Roberts recognize that different levels of effort do not correspond linearly with different levels of efficient work. Therefore, the effective effort rate is a function $e(u(t))$ of $u(t)$, and the work accomplished by time $t$ is $X(t) = \int_0^t e(u(s))\, ds$. The probability of success by time $t$ is now a function $F(X(t))$ of $X(t)$.

The value of a project is influenced not only by the time at which it is completed, it is also influenced by those products the competition develop. In the pharmaceutical industry the most important of these outside influences is the possibility of preemption by a competitor's patent. It is assumed that preemption at time $t$

will make an uncompleted project worthless. The probability of preemption by time t is assumed to be of the form $1 - e^{-\phi t}$, where $\phi$ is estimated by management.

Combining the foregoing results, the expected gross present value of a project with effort u(t) is clearly

$$G(u) = \int_0^\infty V(t)e^{-\phi t}\, dF(X(t))$$

To obtain the expected net present value, the expected cost of the expended effort must be subtracted. As a function of u, the expected discounted effort is defined to be

$$E(u) = \int_0^\infty u(t)[1 - F(X(t)]e^{-\phi t}e^{-Kt}\, dt$$

The factor $[1 - F(X(t))]e^{-\phi t}$ is simply the probability that the allocated effort u(t) at time t is actually used; that is, the project is neither completed (with prob$[1 - F(X(t))]$) nor preempted (with prob $e^{-\phi t}$). In addition, $e^{-Kt}$ is a discount factor influencing the value of the resources being expended at time t. The cost of E(u) is computed internally in the program as the shadow price of these resources in terms of the cash flows that could be generated if they were applied to other projects: specifically, the other projects selected in the plan. While, in fact, a complicated function of the plan being used, this cost will simply be denoted as c. This internal computation of c frees management from having to assign an explicit value to the resources being expended; this is sometimes an advantage, since it may be difficult to determine a monetary value.

Given the function G(u), the marginal rate of gross return at time zero due to a small change u*(s) − u(s) in the discounted effort allocated in the interval (t, t + $\delta$t) may be approximated by the following first-order change in G(u):

$$\Delta \frac{de(u(t))}{du} \int_t^\infty \frac{dF(X(s))}{dX}\left[-\frac{dV(s)}{ds} + \phi V(s)\right]e^{-\phi s}\, ds$$

where $\Delta = \int_t^{t+\delta t}[u*(s) - u(s)]ds$. The terms involving V arise because an increased allocation at time t makes it more likely that successful completion would occur rather than preemption. To the above, one should add a term

$$\int_t^\infty cu(s)e^{-Ks}\, ds$$

because earlier completion frees some effort that would otherwise have been required on the project, and which now may earn c per unit released.

To obtain the expression for the first-order change in the expected net present value at time t (rather than time zero) conditional on the project being still in progress at time t, we divide by the probability of continuation up to time t, and adjust the previous discounting by dividing by the factor $e^{-Kt}$. The corresponding rate of change defines the marginal profitability index

$$
MPI(t) = [1 - F(X(t))]^{-1} \frac{de(u(t))}{du} \int_t^\infty \frac{dF(X(s))}{dX} \left[ \frac{-dV(s)}{ds} \right.
$$

$$
\left. + cu(s)^{-Ks} \right] e^{Kt} e^{-\phi(s-t)} \, ds
$$

By definition of c, it follows that the expected net present value of this project is increased by additional effort at time t if and only if $MPI(t) > c$.

The forms of the functions $V(t)$, $e(u)$, and $F(X)$ are parametric, the user supplying the appropriate parameters. For example, $V(t) = Ve^{-\gamma t}$, where the discounting factor $\gamma$ is broken down into two parts: $\gamma_1 = \log_e(1 + r_1)$, where $r_1$ is the appropriate interest rate (expressed in terms of inflation-free money) for the capital employed by the firm, and $\gamma_2 = \log_e(1 + r_2)$, where $r_2$ defines the additional deterioration in value due to obsolescence. Similarly, $e(u)$ is fitted to be of the form

$$
e(u) = \frac{1 - \exp(-hu)}{L + m\exp(-hu)}
$$

using the manager's estimates of this function at four different values of u.

The function $F(X)$ is more interesting. It essentially depends on an unknown discovery rate $\alpha$ for the project. By means of a prior distribution, which management is aided in specifying, the internally estimated $\alpha$ changes with time, reflecting the fact that prolonged failure implies decreasing likelihood of success. The function $F(X)$ also incorporates another aspect of research. In the initial phases of a project, a large proportion of the resources is expended in acquiring expertise, with this proportion decreasing in time. In the Gittins-Roberts model it is assumed that an efficient level of expertise is approached exponentially at a rate estimated by the research manager. This implies that the initial resource expenditures are not as effective in making discoveries as are those expended later. For a constant rate of effort and a fixed $\alpha$, the probability of successful completion of the project within a time interval $(t, t + \delta t)$, conditional on the project still being under way, increases with t toward a plateau rate.

The Gittins-Roberts model may be illustrated by an example. For each of five projects the information in Table 13 is given together with some details specifying the rate with which expertise is acquired and the effective rate of working. Some initial allocation plan is specified; say, 1, 1, 2, 4, and 4 senior scientists are assigned to the respective projects until their respective completion times. In return, the model gives the information in Table 14 for each time period included in the life of the projects up to some prespecified time horizon. The output also tells us that the marginal expected rate of return from additional effort for this plan is 12.4. This is less than the overall expected return per unit effort, which is 17.7. So were less effort employed, an improvement in research productivity could be made. However, rather than doing this it is perhaps more likely that the manager may wish to explore different allocations between the projects. For this purpose the marginal profitability indices calculated for each project are useful. These MPIs for the first plan are shown in Figure 38. As they define the rate at which the expected present value of the research program increases as small increases are made for the effort allocated to the project, it is clear that it is advantageous to transfer some resources from projects 4 and 5 to projects 2 and 3. This suggests a possible allocation of say, 1, 3, 3, 3, and 2 units to the five projects, respectively. This new plan leads to a total expected discounted return of 393.3 (com-

Table 13  Project Input Information

|  | Project | | | | |
|---|---|---|---|---|---|
|  | 1 | 2 | 3 | 4 | 5 |
| Long-run rate of return on capital | | | 5% | | |
| Present value on successful completion | 50 | 75 | 100 | 125 | 150 |
| Obsolescence rate | 10% | 7.5% | 3.7% | 2.5% | 0% |
| Preemption rate | 1% | 2% | 4% | 5% | 8% |

Table 14   Project States After One Year

|  | Project | | | | |
| Probability of: | 1 | 2 | 3 | 4 | 5 |
| --- | --- | --- | --- | --- | --- |
| Successful completion | 0.04 | 0.15 | 0.34 | 0.35 | 0.54 |
| Preemption | 0.01 | 0.02 | 0.03 | 0.04 | 0.06 |
| Still continuing | 0.95 | 0.83 | 0.63 | 0.61 | 0.40 |

pared to 369.5 for the first plan) for the same expenditure of discounted resources. Once again the research manager has at his or her disposal the corresponding set of MPIs, and so is able to seek further improvements.

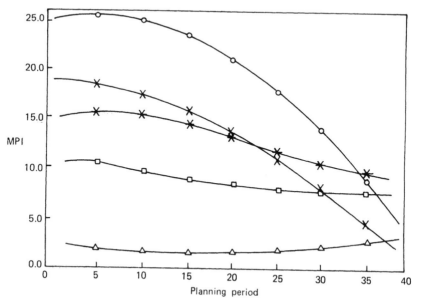

Figure 38   Marginal profitability indices for Plan 1.

## 4.10. MULTIOBJECTIVE MODELS

The models discussed in the previous sections seek to maximize a single objective. Often the purpose of research is to create a product that fulfills, in a satisfactory manner, several objectives, some of which may be conflicting. For example, two such objectives may be maximizing profits and maximizing sales, or, maximizing drug activity and minimizing drug toxicity. A number of approaches exist for handling multiobjective criteria. Essentially, all the methods are seeking a good compromise rather than an optimal solution. Gear (1974) categorizes the available methods as belonging to one of the following types:

1. Objective ordering
2. Indifference surface determination
3. Cost/benefit analysis
4. Utility maximization

The first three may be summarized briefly. The objective ordering method involves the listing of the objectives in order of decreasing importance/relevance. An optimal solution is sought for the first objective. If this solution is also satisfactory with respect to the other, this becomes the chosen policy. If not, the next objective is maximized independently of the previous objective and the result rated with respect to the other objective. This process is repeated until (hopefully) a good solution is obtained. The indifference surface method has been discussed in Section 4.6. There the desirability of low risk was introduced as a constraint, while financial return was maximized subject to this constraint. This technique may be generalized by introducing constraints to represent satisfactory levels of performance for all objectives other than the most important, which is maximized. By altering the degree to which the constraints operate in a systematic way, a family of feasible solutions is generated. Final selection from this set is then left to the decision maker. The last of these three methods is cost/benefit analysis. Here all the costs and benefits associated with the objectives are converted into money terms using unit consts. Then the resulting monetary function is maximized.

The utility maximization approach is similar to the cost/benefit approach. To each possible outcome $(X_1, \ldots, X_n)$ of a portfolio R, where $X_i$ is the benefit measured in appropriate units obtained with respect to objective i, a utility $U(X_1, \ldots, X_n)$ is assigned. This assignment must satisfy a set of axioms guaranteeing consistency in our preferences. For example, if outcome

$(X_1, \ldots, X_n)$ is preferred to $(X'_1, \ldots, X'_n)$, and $(X'_1, \ldots, X'_n)$ is preferred to $(X''_1, \ldots, X''_n)$, we must have $U(X_1, \ldots, X_n) > U(X'_1, \ldots, X'_n) > U(X''_1, \ldots, X''_n)$. Details regarding thses axioms may be found in Fishburn (1970) or DeGroot (1970). It may be shown that under uncertainty a decision maker behaves rationally with respect to these axioms if he or she defines the objective to be one of maximizing expected utility $E_R(U(X_1, \ldots, X_n))$, where the maximum is taken over all possible portfolios R. In this fashion, a multiobjective allocation problem has been reduced to a single objective allocation problem.

The difficulty with the utility approach is always one of specifying the utility function U. This requires that management be able to examine explicitly the trade-offs appropriate between all possible benefit outcomes $(X_1, \ldots, X_n)$. An interesting article by Keefer (1978) examines one selection problem where this was done successfully. For each of six different objectives a marginal utility function $U_i(X_i)$ was constructed for each of the objectives, where $U_i(X_i)$ defines the utility for outcome $X_i$ alone. Typically, such a function is concave, exhibiting decreasing marginal returns. Details of the construction are provided by Keefer. Then the research managers were questioned about the relative independence of the different objectives. Specifically, let

$$\bar{X}_i = (X_i, \ldots, X_{i-1}, X_{i+1}, \ldots, X_n)$$

$$\bar{X}_{ij} = (X_1, X_2, \ldots, X_{i-1}, X_{i+1}, \ldots, X_{j-1}, X_{j+1}, \ldots, X_n)$$

Then $X_i$ is said to be *utility independent* of $\bar{X}_i$ if preference for risky choices over $X_i$ with the value of $\bar{X}_i$ held fixed do not depend on the fixed value of $\bar{X}_i$. The set $\{X_i, X_j\}$ is *preferentially independent* of $\bar{X}_{ij}$ if preferences for consequences differing only in the values of $X_i$ and $X_j$ do not depend on the fixed value of $\bar{X}_{ij}$. The information elicited from the research managers showed that these conditions were satisfied. By a theorem of Keeney (1974), it followed that a manager's utility function $U(X_1, \ldots, X_n)$ was necessarily of the form

$$U(X_1, \ldots, X_n) = \left[ \prod_{i=1}^{n} (1 + KR_i U_i(X_i)) - 1 \right] K^{-1}$$

where K and $R_i$ are appropriately chosen constants. The expected value of this function was maximized over the set of possible portfolios. The research managers found the solution interesting and the technique useful.

The "best compromise" method has been explored independently ly by Freeman (1972) and Benayoun et al. (1971). It is a form of linear programming in which a regret function is minimized. It may be formulated as follows. Let $X_i$ be the variable representing project i, and for each project define

$c_{ik}$ = resource requirement of project i for resource k

$b_{ij}$ = benefit of selecting project i with respect to criterion j

Also define

$W_j$ = weight representing the importance of objective j

$C_k$ = constraint on resource k

For each criterion j = 1,...,J solve the integer programming problem that maximizes the benefits with respect to criterion j; that is, solve

$$\max \sum_{i=1}^{N} X_i b_{ij}$$

subject to the constraints

$$\sum_{i=1}^{N} X_i c_{ik} \leqslant C_k, \quad k = 1,...,K$$

Let the value of the associated optimal portfolio be $B_j^*$. Now define for any portfolio the vector $(B_1,...,B_J)$ of values of this portfolio with respect to each of the criteria, and associate with it a vector of regrets $(\Delta B_1,...,\Delta B_J)$, where $\Delta B_j = (B_j^* - B_j)/B_j^*$.

Our objective now is to choose a portfolio such that $\Delta B_j$ is as close to zero as possible for each j = 1,...,J. One possible form of this objective is the minimization of the maximum weighted regret; that is,

$$\text{minimize} \max_{1 \leqslant j \leqslant J} \{\Delta B_j W_j\}$$

subject to the constraints

$$\sum_{i=1}^{N} X_i c_{ik} \leqslant C_k, \quad k = 1,...,K$$

Another alternative is to minimize the weighted sum of the regrets, that is,

$$\text{minimize} \sum_{j=1}^{J} \Delta B_j W_j$$

subject to the same constraints.

## REFERENCES

Aaker, A., and Tyebjee, T. T. (1978), "A model for the selection of interdependent R&D projects" *IEEE Trans. Eng. Manage.*, Vol. EM-25, No. 2, pp. 30–36.

Adams, B. H., and Gearing, C. E. (1974), "Determining an optimal set of research experiments." *IEEE Trans. Eng. Manage.*, Vol. EM-21, No. 1, pp. 20–39.

Albala, A. (1975), "A stage approach for the evaluation and selection of R&D projects," *IEEE Trans. Eng. Manage.*, Vol. Em-22, No. 4, pp. 153–164.

Allen, D. H., and Johnson, T. F. N. (1970), "Optimal Selection of a research project portfolio under uncertainty," *Chem. Eng.*, No. 241, pp. 278–284.

Allen, D. H., and November, P. J. (1969), "Optimum information accumulation patterns in the development of new chemicals," *Chem. Eng.*, No. 234, pp. 429–435.

Ansoff, H. I. (1964), "Evaluation of applied R&D in a form," in *Technological Planning in the Corporate Level,* J. R. Bright, ed., Harvard University Press, Cambridge, Mass.

Asher, D. T. (1962), "A linear programming model for the allocation of R&D Efforts," *IRE Trans. Eng. Manage.*, Vol. EM-3, pp. 154–157.

Atkinson, A., and Bobis, A. (1969), "A mathematical basis for the selection of research projects," *IEEE Trans. Eng. Manage.*, Vol. EM-16, No. 1, pp. 2–8.

Augood, D. R. (1973), "A review of R&D evaluation methods," *IEEE Trans. Eng. Manage.*, Vol. EM-20, No. 4, pp. 114–120.

Augood, D. R. (1975), "A new approach to R&D evaluation," *IEEE Trans. Eng. Manage.*, Vol. EM-22, No. 1, pp. 2–9.

Baker, N. R. (1974), "R&D project selection models: an assessment," *IEEE Trans. Eng. Manage.*, Vol. EM-21, pp. 165–171.

Baker, N. R., and Reeland, J. (1975), "Recent advances in R&D benefit measurement and project selection methods," *Manage. Sci.*, Vol. 21, No. 10, pp. 1164–1175.

Baker, N. R., and Pound, W. H. (1964), "R&D project selection: where we stand," *IEEE Trans. Eng. Manage.*, Vol. EM-11, pp. 124–134.

Beged-Dov, A. G. (1965), "Optimal assignment of R&D projects in a large company using an integer programming model," *IEEE Trans. Eng. Manage.*, Vol. EM-12, pp. 138–142.

Bell, D. C. (1969), *The Evaluation and Selection of Research Projects*, British Gas Council, Operational Research Dept., Report 35, London.

Bell, D. C., and Read, A. W. (1970), "The application of a research project selection method," *R&D Manage.*, Vol. 1, No. 1, pp. 35-42.

Bellman, R. E. (1957), *Dynamic Programming*, Princeton University Press, Princeton, N.J.

Benayoun, R., de Montgolfier, J., Tergny, J., and Laritchev, O. (1971), "Linear programming with multiple objectives: step method (STEM)," *Math. Programming*, Vol. 1, pp. 367–375.

Bender, A. D., and Pyle, E. B. (1972), "Planning, control and resource allocation models in R&D: a performance review by a pharmaceutical manufacturer," *XIX International Meeting of TIMS*, Apr., Houston, Texas.

Bergman, S. W. (1981), "*Acceptance sampling: the buyer's problem,*" Ph.D. dissertation, Yale University.

Bradbury, (1973), "Qualitative aspects of the evaluation and control of research and development projects," *R&D Manage.*, Vol. 3, No. 2, pp. 49–57.

Brown, R. (1978), "Probabilistic models of project management with design implications," *IEEE Trans. Eng. Manage.*, Vol. EM-25, No. 2, pp. 43–49.

Brunings, K. (1976), "Decline of pharmaceutical research and development," *Res. Manage.*, Vol. 19, No. 6, p. 19–22.

Brunings, K. (1979), "The role of basic research in development of medicinal products," *Res. Manage.*, Vol. 22, No. 4, pp. 19–23.

Buel, W. D. (1960), "A simplification of Hay's method of recording paired comparisons," *J. Appl. Phychol.*, Vol. 44, pp. 347–348.

Cetron, J., Martino, J., and Roepke, L. (1967), "The selection of R&D program content—survey of quantitative methods," *IEEE Trans. Eng. Manage.*, Vol. EM-14, No. 1, pp. 4–13.

Charnes, A., and Cooper, W. W. (1963), "Deterministic equivalents for optimizing and satisfying under change constraints," *Oper. Res.*, Vol. II, No. 1, pp. 18–39.

Chiu, L., and Gear, T. E. (1979), "An application and case history of a dynamic R&D portfolio selection model," *IEEE Trans. Eng. Manage.*, Vol. EM-26, No. 1, pp. 2−7.

Churchman, C. W., Ackoff, L., and Arnoff, E. L. (1957), *Introduction to Operations Research*, Wiley, New York.

Clark, P. (1977), "A profitability project selection method," *Res. Manage.*, Vol. 20, No. 6, pp. 29−33.

Clarke, T. E. (1976), *R&D Management Bibliography*, The Innovation Management Institute of Canada.

Clymer, H. A. (1970), "The changing costs and risks of innovation in drug development," *Res. Manage.*, Vol. 13, No. 5, pp. 375−382.

Cochran, M. A., Pyle, E. B., Greene, L. C., Clymer, H. A., and Douglas, A. B. (1971), "Investment model for R&D project evaluation and selection," *IEEE Trans. Eng. Manage.*, Vol. EM-18, No. 3, pp. 89−99.

Cooper, J. M. (1978), "An evaluation system for project selection," *Res. Manage.*, Vol. 21, No. 4, pp. 29−35.

Cox, J. S. G., and Styles, A. E. J. (1979), "From lead compound to product," *R&D Manage.*, Vol. 9, No. 3, pp. 125−135.

Cox, J. S. G., Millane, B. V., and Styles, A. E. J. (1975), "A planning model of pharmaceutical research and development," *R&D Manage.*, Vol. 5, No. 3, pp. 219−227.

Davies, O. L. (1962), "Some statistical considerations in the selection of research projects in the pharmaceutical industry," *Appl. Statist.*, Vol. 11, pp. 170−183.

Davies, C. (1971), "Application of systems engineering techniques to projects in the chemical process industry," *Chem. Eng.* (London), No. 248, (Apr.), pp. 149−152.

Dean, B. V. (1968), *Evaluating, Selecting, and Controlling R&D Projects*, American Management Association, Inc., New York.

Dean, B. V., and Goldhar, J. L., eds. (1980), *Management of Research and Innovation*, North-Holland, New York.

Dean, B. V., and Hauser, L. E. (1967), "Advanced material systems planning," *IEEE Trans. Eng. Manage.*, Vol. EM-14, No. 1, pp. 21−43.

Dean, V. B., and Nishry, M. J. (1965), "Scoring and profitability models for evaluating and selecting engineering projects," *Oper. Res.*, Vol. 13, pp. 550−569.

Dean, V. B., and Roepcke, L. (1969), "Cost effectiveness in R&D resource allocation," *IEEE Trans. Eng. Manage.*, Vol. EM-16, No. 4, pp. 222−242.

DeGroot, M. H. (1970), *Optimal Statistical Decisions*, McGraw-Hill, New York.

Deshmukh, S. D., and Chikte, S. D. (1980), "A unified approach for modeling and analyzing new project R&D decisions," *Research and Innovation*, B. V. Dean and J. L. Goldhar, North-Holland, Amsterdam, pp. 163–182.

Disman, S. (1962), "Selecting R&D projects for profit," *Chem. Eng.*, Vol. 11, Dec., pp. 87–90.

Ebert, R. J. (1970), "Methodology for improving subjective R&D estimates," *IEEE Trans. Eng. Manage.*, Vol. EM-17, No. 3, pp. 108–115.

Edwards, C. C. (1974), "The role of government and FDA regulations in drug R&D," *Res. Manage.*, Vol. 17, No. 2, pp. 21–23.

Faust, R. E. (1971), "Project selection in the pharmaceutical industry," *Res. Manage.*, Vol. 14, No. 5, pp. 46–50.

Faust, R. E. (1975), "Pharmaceutical research planning strategies," *J. Soc. Res. Admin.*, Vol. VII, No. 3.

Faust, R., and Ackerman, G. (1974), "Program/project management at Hoffman-LaRoche," *Res. Manage.*, Vol. 17, No. 1, pp. 38–42.

Fishburn, P. C. (1970), *Utility Theory for Decision Making*, Wiley, New York.

Flin, H., and Turbin, A. S. (1970), "Decision through analysis for industrial research," *Res. Manage.*, Vol. 13, No. 1, pp. 27–30.

Freeman, R. J. (1960), "A stochastic model for determining the size and allocation of the research budget," *IRE Trans. Eng. Manage.*, Vol. EM-7, No. 1, pp. 2–7.

Freeman, P. (1972), "Development of linear programming models for portfolio selection in research and development," Ph.D. dissertation, Manchester University.

Freeman, P., and Gear, A. E. (1971), "A probabilistic objective function R&D portfolio selection," *Oper. Res. Quart.*, Vol. 22, pp. 253–265.

Gargiulo, C. R., Hannoch, J., Hertz, D. B., and Zang, T. (1961), "Developing systematic procedures for directing research programs," *IRE Trans. Eng. Manage.*, Vol. EM-8, No. 1, pp. 24–29.

Gauthier, G. G., et al. (1977), "Professional employees in drug R&D: Personal characteristics and work time patterns, Part I," *J. Soc. Res. Adminis.*, Vol. VIII, No. 4, pp. 9–16.

Gear, A. E. (1974), "A review of some recent developments in portfolio modeling in applied research and development," *IEEE Trans. Eng. Manage.*, Vol. EM-21, No. 4, pp. 119–125.

Gear, A. E., and Lockett, A. G. (1973), "A dynamic model of some multi-stage aspects of research and development portfolios," *IEEE Trans. Eng. Manage.*, Vol. EM-20, No. 1, pp. 22—28.

Gear, A. E., and Lockett, A. G. (1974), "A dynamic model of some multi-stage aspects of research and development portfolios," *IEEE Trans. Eng. Manage.*, Vol. EM-21, No. 4, pp. 141—147.

Gear, A. E., Lockett, A. G., and Pearson, A. W. (1971), "Analysis of some portfolio selection models for R&D," *IEEE Trans. Eng. Manage.*, Vol. EM-18, No. 2, pp. 66—75.

Geoffrion, A. N., et al. (1972), "An interactive approach for multi criterion optimization with an application to the operation of any academic department," *Manage. Sci.*, Vol. 19, No. 4, pp. 357—368.

Gillespie, J. S., and Gear, A. E. (1973), "An analytical methodology for comparing the suitability of management science models," *IEEE Trans. Eng. Manage.*, Vol. EM-20, No. 4, pp. 121—129.

Gittins, J. C. (1973), "Speculative chemical research—How many eggs in a basket?" *R&D Manage.*, Vol. 3, No. 2, pp. 71—81.

Gittins, J. C. (1979), "Bandit processes and dynamic allocation indices," *J. Statist. Soc. B,* Vol. 41, pp. 148—177.

Gittins, J. C., and Jones, D. M. (1974a), "A dynamic allocation index for new-product chemical research," Department of Engineering Technical Report, University of Cambridge, Cambridge.

Gittins, J. C., and Jones, D. M. (1974b), "A dynamic allocation index for the sequential design of experiments," *Proc. Eur. Meet. Statist.*, Hungarian Academy of Sciences, Budapest, 1972, ed., J. Gani, North Holland, Amsterdam, pp. 725—736.

Gittins, J. C., and Roberts, D. M. (1981), "RESPRO—an interactive planning procedure for new product chemical research," *R&D Manage.*, Vol. 11, No. 4, pp. 139—148.

Glazebrook, K. D. (1976), "A profitability index for alternative research projects," *OMEGA, Int. J. Manage. Sci.*, Vol. 4, No. 1, pp. 79—83.

Glazebrook, K. D., (1977), "Stochastic scheduling with order constraints," *Int. J. Syst. Sci.*, Vol. 7, pp. 657—666.

Glazebrook, K. D. (1978), "Some ranking formulae for alternative research projects," *OMEGA, Int. J. Manage. Sci.*, Vol. 6, No. 2, pp. 193—194.

Gloskey, C. R. (1960), "Research on a research department: an analysis of economic decisions on projects," *IRE Trans. Eng. Manage.*, Vol. 7, No. 4, pp. 166—172.

Hall, M. (1964), "The determinants of investment variations in research and development," *IEEE Trans. Eng. Manage.*

Harris, J. S. (1961), "New project profile chart," *Chem. Eng. News*, Vol. 39, Apr. 17, pp. 110–118.

Hart, A. (1966). "A chart for evaluating product research and development projects," *Oper. Res. Quart.*, Vol. 17, No. 4, pp. 347–358.

Hess, S. W. (1962), "A dynamic programming approach of R&D budgeting and project selection," *IRE Trans. Eng. Manage.*, Vol. EM-9, Dec., pp. 170–179.

Jarvis, P. E. J., and Rippin, D. W. T. (1976), "A company model for research and development," *R&D Manage.*, Vol. 6, No. 3, pp. 115–123.

Kadane, J., and Simon, H. (1977), "Optimal strategies for a class of constrainted sequential problems," *Ann. Statist.*, Vol. 5, No. 2, pp. 237–255.

Keefer, D. L. (1976), "A decision analysis approach to resource allocation planning problems with multiple objectives," Ph.D dissertation, Dept. of Industrial and Operations Engineering, University of Michigan, Ann Arbor, Mich.; available from University microfilms, Ann Arbor.

Keefer, D. L. (1978), "Allocation planning for R&D with uncertainty and multiple objectives" *IEEE Trans. Eng. Manage.*, Vol. EM-25, No. 1, pp. 8–14.

Keeney, R. L. (1974), "Multiplicative utility functions," *Oper. Res.*, Vol. 22, pp. 22–34.

Kepler, C. E., and Blackman, A. W. (1973), "The used dynamic programming techniques for determining resource allocations among R/D projects: an example," *IEEE Trans. Eng. Manage.*, Vol. EM-20, No. 1, pp. 2–5.

Lockett, A. G., and Freeman, P. (1970), "Probabilistic networks and R&D portfolio selection," *Oper. Res. Quart.*, Vol. 21, No. 3, pp. 353–360.

Lockett, A. G., and Gear, A. E. (1972), "Programme selection in research and development," *Manage. Sci.*, Vol. 18, No. 10, pp. B575–B590.

Lockett, A. G., and Gear, A. E. (1973), "Representation and analysis of multi-stage proglems in R&D," *Manage. Sci.*, Vol. 19, No. 8, pp. 947–960.

Maher, P. M. (1973), "Attitudes and conclusions resulting from an experiment with computer-based R&D project selection techniques," *R&D Manage.*, Vol. 4, No. 1, pp. 1–8.

Minkes, A. L., and Samuels, T. M. (1966), "Allocation of research and development expenditures in the firm," *J. Manage, Stud.*, Vol. 3, pp. 62–72.

Moore, J., and Baker, N. (1969a), "An analytical approach to scoring model design—application to research and development projection selection," *IEEE Trans. Eng. Manage.*, Vol. EM-16, No. 3, pp. 90—98.

Moore, J., and Baker, N. (1969b), "Computational analysis of scoring models for R&D project selection," *Manage. Sci.*, Vol. 16, No. 4, pp. B 212—B 232.

Mottley, C. M., and Newton, R. D. (1959), "The selection of projects for industrial research," *Oper. Res.*, Vol. 7, No. 6, pp. 740—751.

National Economic Development Office (1972), *Focus on Pharmaceuticals: A Report by the Pharmaceuticals Working Party of the Chemicals EDC*, H. M. Stationery Office, London.

Nutt, A. B. (1965), "An approach to research and development effectiveness," *IEEE Trans. Eng. Manage.*, Vol. EM-10, No. 3, pp. 103—112.

Osbaldeston, M. D., Cox, J. A., and Loveday, D. E. (1978), "Creativity and organization in pharmaceuticals R&D," *R&D Manage.*, Vol. 8, No. 3, pp. 165—175.

Paolini, A., and Glaser, M. (1977), "Project selection methods that pick winners," *Res. Manage.*, Vol. 20, No. 3, pp. 26—29.

Pappas, C. F., and MacLaren, D. D. (1961), "An approach to research planning," *Chem. Eng. Prog.*, Vol. 57, No. 5, pp. 65—69.

Pearson, A. W. (1972), "The use of ranking formulae in R&D projects," *R&D Manage.*, Vol. 2, No. 2, pp. 69—73.

Pound, W. H. (1964), "Research project selection: testing a model in the field," *IEEE Trans. Eng. Manage.*, Vol. EM-11, No. 1, pp. 16—22.

Raiffa, H. (1968), *Decision Analysis: Introductory Lectures on Choices Under Uncertainty*, Addison-Wesley, Reading, Mass.

Reader, R. D., James, M. F., and Goodman, R. W. (1966), "The evaluation and selection of research projects—a progress report," *British Iron and Steel Research Association (U.K.)*, O.R. Department, Ref. No. OR/43/66.

Reichner, A. (1967), "The inclusion of unforeseen occurrences in decision analysis," *IEEE Trans. Eng. Manage.*, Vol. EM-14, No. 4, pp. 177—182.

Reis-Arndt, E., and Elners D. (1972), "Results of pharmaceutical research: new pharmaceutical agents, 1961—1970," *Drugs Made Ger.*, Vol. 15, pp. 134—140.

Roberts, K., and Weitzman, L. (1980), "On a general approach to search and information gathering," Working Paper 263, Dept. of Economics, MIT.

Rosen, E. M., and Souder, W. E. (1965), "A method for allocating R&D expenditures," *IEEE Trans. Eng. Manage.*, Vol. EM-12, No. 3, pp. 87—93.

Sarett, L. (1974), "FDA regulations and their influence on future R&D," *Res. Manage.*, Vol. 17, No. 2, pp. 18—20.

Savage, L. J. (1970), "The elicitation of personal probabilities," Report, Department of Statistics, Yale University, New Haven, Conn.

Savage, L. J. (1972), The *Foundations of Statistics*, Dover, New York.

Sengupta, S. S., and Dean, B. V. (1960), "On a method of determining corporate research and development budgets," *Management Science Models and Techniques*, C. W. Churchman and M. Verhulst, eds., Pergamon Press, Elmsford, N.Y., pp. 210—225.

Sigford, W. E., et al. (1965), "A methodology for determining relevance in complex decision making," *IEEE Trans. Eng. Manage.*, Vol. EM-12, No. 1, pp. 9—13.

Sobin, D. (1965), "Proposal generation and evaluation methods in research and exploratory development," Research Analysis Corporation Report RAC-R-11, Roanoke, Va.

Souder, W. E. (1966), *Operations Research in R&D*, The Monsanto Company, St. Louis, Mo.

Souder, W. E. (1972a), "A scoring methodology for assessing the suitability of management science models," *Manage, Sci.*, Vol. 18, No. 10, pp. B526—B543.

Souder, W. E. (1972b), "Comparative analysis of R&D investment models," *AIIE Trans.*, Vol. 4, No. 1, pp. 57—64.

Souder, W. E. (1973), "Analytical effectiveness of mathematical models for R&D project selection," *Manage, Sci.*, Vol. 19, No. 9, pp. 907—913.

Souder, W. E. (1977), "Effectiveness of nominal and interacting group decision process for integrating R&D and marketing," *Manage, Sci.*, Vol. 23, No. 6, pp. 595—605.

Souder, W. E. (1978), "A system for using R&D project evaluation methods," *Res. Manage.*, Vol. 21, No. 5, pp. 29—37,

Souder, W. E., Maher, P. M., Shumway, C. R., Baker, N. R., and Rubenstein, A. H. (1974), "Methodology for increasing the adoption of R&D project selection models," *R&D Manage.*, Vol. 4, No. 2, pp. 75—83.

Stucki, J. C. (1980), "A goal-oriented pharmaceutical research and development organization: an eleven-year experience," *R&D Manage.*, Vol. 10, No. 3, pp. 97—105.

Styles, A., and Cox, J. S. (1977), "Balancing Resources in pharmaceutical research," *R&D Manage.*, Vol. 8, No. 1, pp. 1−12.

Tukey, T. W. (1977), *Exploratory Data Analysis*, Addison-Wesley, Reading, Mass.

Van Roy, T. J., and Gilders, L. F. (1978), "A practical tool for improved resource allocation: the dynamic time new procedure," *IEEE Trans. Eng. Manag.*, Vol. EM-25, No. 4, pp. 93−97.

Walters, L. D. (1967), *R&D project selection: interdependence and multiperiod probabilistic budget constraints*, Ph.D thesis, Arizona State Univeristy, Tempe.

Weatherall, M. (1972), "Change and choice in the discovery of drugs," *Proc. R. Soc. Med.*, Vol. 65, pp. 329−334.

Weitzman, M. L. (1979), "Optimal search for the best alternative," *Econometrics*, Vol. 47, No. 3, pp. 641−655.

Winkofsky, E. P., Mason, R. M., and Souder, W. E. (1980), "R&D budgeting and project selection: a review of practices and models," *Management of Research and Innovation*, B. V. Dean and J. V. Goldhar, eds., North-Holland, Amsterdam, pp. 183−197.

Winter, O. (1969), "Preliminary economic evaluation of chemical processes at the research level," *Industrial Eng. and Chem.*, Vol. 61, No. 4, pp. 45−52.

# SUMMARY: STATE OF THE ART

One of the companies that helped us gather the material for this book judges the suitability of its employees for eventual promotion to senior positions partly on the basis of what it calls "helicopter." By this is meant the ability to see the company, its strengths and weaknesses, its opportunities and position in relation to competitors, in a true perspective, uncluttered by sectional interests or short-term considerations: to have, in brief, a clear bird's-eye view of the situation. This chapter is an attempt at a similar assessment of statistical methods as an aid in pharmaceutical and agrochemical research. Brief comments follow on the state of the art in the areas covered by the four chapters of this book. We should like at the outset to reiterate our general impression that the methods available are often better than their neglect in practice might suggest.

The rise of QSAR methods of a statistical nature over the past decade or so has coincided with considerable progress in the pharmacologist's understanding of the way drugs act. Correspondingly, there are two possible emphases that can be given to the search for active compounds, the first concentrating on accumulating and sifting data on large numbers of compounds, and the second on pinpointing the precise compound to be synthesized. The conflict between the two should not be exaggerated, and an efficient research plan will in most cases use both methods in parallel to select compounds for testing. To achieve this, a close working relationship between chemist and statistician is clearly desirable, and this in turn points to the desirability of each having a greater understanding of the other's subject than is common at present. New methods should emerge from such a partnership, making use of advances in chemical knowledge as well as taking uncertainty into account.

For designing a single screen the models described in Chapter 2 go a long way toward providing what is needed. Some of the constrained optimization models have the advantage of avoiding the need to quantify costs and rewards. These models are most appropriate for short-term planning, when there is, for example, not enough time for the supply of mice to be increased. In the longer term such constraints can be removed, at a cost, and a suitable model must balance this against other costs and rewards. Davies's model described in Section 2.6 is appropriate and could be used, although ideally one would like to add some refinements.

One desirable refinement for any sequential design for a single screen is to specify the sequence of dose levels at which compounds are tested. It would not be difficult to arrive at such a sequence, making the standard assumption of a logistic relationship between dose and response. Second, the assumption of a two-point distribution of activity, classifying compounds simply as active or inactive, is very crude. It would be preferable to assume a negative exponential distribution, as observed in practice by Davies, perhaps also allowing a positive probability of zero activity. This, again, should be fairly straightforward.

The organization of a sequence of screens is more complex and, as already mentioned, here the literature seems to have nothing to offer. To make the problem more tractable, some simplifying assumptions are needed. In particular, it is convenient to consider separately the questions of the sequencing of the various activity and toxicity screens, and of the rules for accepting or rejecting compounds at each screen in the resulting sequence. The answers to these questions interact to some extent, but not so much as to prevent one from getting sensible answers by tackling them separately. Models for the two resulting subproblems were presented in Chapter 3.

As far as the sequencing of the screens is concerned, the model of Section 3.2 may well be all that is required. Just possibly one might want to allow the sequence of screens to be different for different compounds, depending on the test results at earlier screens. A model to allow for this could also be solved by dynamic programming. Models of this kind are discussed by Gittins (1982) (see Section 3.3).

Once the sequence of the screens has been settled, rules for the acceptance or rejection of compounds may in principle be determined using a model like that described in Section 3.1. This model extends the single-screen model of Section 2.6, and some further changes are necessary when later screens are included, at which the number of compounds tested may well be very small. At these later stages it is no longer a question of selecting the best of the compounds available, but rather of developing all those compounds for which the expected rewards of doing so exceed the costs. The time scale for the multiscreen model is also much longer than for a single screen, so that future rewards and costs should be discounted. These changes could be incorporated fairly easily.

Greater difficulties arise from the need to make realistic distributional assumptions without this resulting in a computationally intractable model. As for a single screen, it should be possible to make some progress by assuming a negative exponential distribution of activity (or toxicity), with a positive probability at zero. A suitable joint distribution for the activity exhibited at two successive screens, or for activity and toxicity, could be chosen so that the marginal distributions were as just stated. The need for computational tractability might well prove to limit the possibilities to the simultaneous design of two successive screens, and this may be sufficient for practical purposes. It would, for example, reveal the point at which it is better to pass a compound on for testing at a screen which, although more expensive, is more likely to reflect the activity in humans. A sequence of screens could be designed by considering successive screens two at a time, temporarily regarding earlier screens in the sequence as fixed, determining the prior distribution of activity and toxicity for compounds tested by the first screen of the pair, and later screens also as fixed, determining the expected net reward as a function of activity and toxicity for compounds passed by the second screen of the pair. Such a scheme was discussed in Section 3.1.

In the field of project selection the various checklists, and profitability and other indices, which have been proposed are undoubtedly of some value. Most of the more sophisticated models, however, are not specifically designed for new-product chemical research and are insufficiently flexible to cope with its enormous uncertainties. The RESPRO procedure described in Section 4.9 attempts to fill this gap, but it will certainly not do so completely. For example, it should be possible to devise a model that allows the research manager to explore the possibilities of different combinations of high-risk and potentially very profitable projects, and low-risk projects with less potential for high profits. Another model might be based on research areas rather than individual projects. At the most macroscopic level such a model might be used to suggest how the total research effort of a laboratory should be divided between pharmaceutical and agrochemical projects. A model of this kind would be particularly appropriate if scientific personnel are not readily transferable between research areas.

Finally, it is worth noting that the various procedures reviewed in this book are appropriate at different levels in the administrative structure of the laboratory. The selection of compounds for testing, the QSAR methods designed for this purpose, and the design of individual screens all fall within the scope of a single project. The necessary decisions at this level may for the most part be taken by the senior scientist concerned. The sequencing of different screens, and the balance of effort between screens, while also matters involving just one project at a time, may have implications for the allocation of resources between different departments, and therefore require decisions at a higher level and on a longer-term basis. The most obvious example here is the deliberate distinction in many laboratories between the functions of testing compounds for activity, and for toxicity or other negative side effects. Project selection, and the allocation of resources between projects, are clearly questions for laboratory management, and are likely to be considered quarterly, or even less frequently. Questions of allocation between research areas are matters for decision by the board of directors, and the time scale is measured in years.

# INDEX

Milton Keynes UK
Ingram Content Group UK Ltd.
UKHW040108071024
449327UK00019B/910